Seed Production of Commercial Vegetables

Seed Production of Commercial Vegetables

Y.S. Sreenivas

Oxford Book Company
Jaipur ▪ New Delhi

ISBN : 9788189473662

Published in India, 2026

Oxford Book Company
Regd. Off. "Bony Residency", Gate No. 2
Opp. Tilak Public School, Vishweswariya Nagar,
Gopalpura Road, Jaipur - 302018, India
Ph: 0141 - 2761280, Telefax: 2761381
email - oxfordbookjpr@gmail.com
website - oxfordbookcompany.com

Delhi Office: 102, 1st Floor, Satyam House,
4327/3, Ansari Road, Darya Ganj,
New Delhi - 110002, India
Ph: 011-45652440

Distributed By
KEYA INTERNATIONAL PRESS LLP
E-mail: keyainternationalpressllp@gmail.com

Printed and bound in India by Thomson Press Ltd., New Delhi

Preface

With changes in dietary habits and consumption of food away from home, the demand for commercial vegetables like potatoes, tomatoes, lettuce, onions and cabbage has risen considerably. The production of these vegetables is highly intensive as compared with the production of such field crops as corn, wheat and sugar beets. The potential crop value is great and justifies the use of comparatively high expenditure on labour, water, fertiliser, farm chemicals, mechanisation and other resources. In such a scenario, it becomes pertinent to consider efficient seed production of commercial vegetables.

The present text is an attempt at producing a manual for vegetable growers on efficient seed production, especially as it differs from conventional crop seed production in the need to maintain strong genetic purity. Guiding readers on seed viability, testing and germination techniques based on specific agroclimatic conditions and growers weeds, it also looks into the lucrative offers of seed production of commercial vegetables especially as the demand for these grow continually. In addition, space has also been given to analysing current trends and developments in the field, as well as making assessment of future trends.

Y.S. Sreenivas

Contents

1

Introduction

Vegetables are harvested for some component of their vegetative (celery, carrot) or reproductive (tomato, broccoli) structure. In most cases, the harvest of the market-use crop requires environmental conditions that are very different from those for reproductive and development of a mature seed crop. For example, ambient temperature, photoperiod, length of growing season, and precipitation amount and distribution may vary for the optimum production of market-use and seed crops.

Ambient temperatures have very different effects on vegetables and their end use. Those vegetables that are harvested for market-use such as carrot, onion, radish, brassicas, and lettuce are typically grown in areas with lower temperatures from those required for a seed crop in order to maintain good vegetative growth and minimise bolting. Seed production of these crops often requires higher temperatures to encourage bolting although extremely high temperatures also hinder pollen growth and fertilisation which lead to reduced seed number per floret or fruit.

This is true in onion, tomato, and lettuce. In contrast, beet and carrot market-use crops have been selected to avoid bolting in mild climates but they still require adequate periods of cool temperatures (vernalisation) for flowering and seed production. The Pacific Northwest of the United States provides such sustained cool temperatures and this is the reason why beet and carrot seed production occurs in this region while market-use production of these crops is geographically more diverse.

The requirement of some vegetables for specific photoperiods before induction of flowers also limits those areas where vegetable seed production is successful. This effect is dependent on the latitude and the length of the photoperiod which greatly affect seed yield. For example, onion is a vegetable crop that exhibits a wide diversity in photoperiod response and this influences where seed production occurs. Seed production of long-day onions occurs at latitudes greater than 45° N and short-day onions at latitudes no greater than 35° N. Limited seed production also can be accomplished in glasshouses where supplemental light is provided or light excluded to promote flowering.

Even when appropriate photoperiods are present, temperatures must be suitable for the completion of the entire life cycle of the crop during the growing season. Successful market-use crops such as onion, radish, broccoli, and cabbage are harvested before flowering occurs. As a result, these crops are grown in regions with cooler temperatures not suitable for seed crops which require longer periods and warmer temperatures for maturation of seed.

Precipitation amount and distribution also varies for market-use and seed crops. Many vegetable crops harvested for their vegetative structure require large amounts of early season rainfall for most rapid development. Seed crops, in contrast, demand uniform precipitation during development followed by dry conditions during seed maturation. Under these environments, dependable conditions for seed harvest are possible and pathogen infestations minimised. This is the principal reason much of the vegetable seed production has shifted to the western United States.

The Mediterranean climate of the Pacific Northwest assures cool, dry summers and warm, wet winters for optimum development of many biennial seed crops. Southern California and much of Arizona are hot and dry during the summer which is ideal for vegetable seed production. The dry condition of these locations permits seed crops to be irrigated with the proper amount of moisture at the proper time so that optimum seed quality is achieved.

VEGITABLE SEED PRODUCTION

The following represents general requirements and considerations utilised in vegetable seed production.

Tillage. The preparation of a soil site for vegetable seed production differs little from that for fresh market production. Soils should have a high water holding capacity to encourage uniform vegetative growth. Depending on the crop and water availability, planting is done on raised, shaped, or flat planting surfaces which require appropriate soil preparation. In all cases, the seed bed must be uniform because most vegetable seed crops are precision planted and uniform emergence and seedling development necessary for optimum performance.

Planting. The planting of vegetable seeds for seed production is generally done by direct drilling or transplanting of glass house grown seedlings. Biennial vegetable seed production, however, can be achieved using two planting techniques. The "seed-to-seed" method is done by planting biennial seed, allowing the resulting plant to overwinter without being transplanted, and harvesting the seed crop the following season. This method does not permit the selection or roguing of root or other genotypic characters. A majority of carrot and onion seed crops are produced by this method which is less expensive and complex than the "root-to-seed" method.

The "root-to-seed" method is done by planting biennial seed, removing the resulting plant from the soil (in some cases stored), replanting, and harvesting the seed crop the following season. The replanted plant is referred to as a "steckling." This method permits the selection or roguing of root or other genotypic characters. It also decreases the time required for varietal development. For example, onion bulbs adapted to dry regions can be planted in the spring in the desert regions of the southwest and then shipped for seed production to central California or the Pacific Northwest for planting in the early fall of the same year. This is the most expensive method of vegetable seed production due to transportation, handling, and replanting costs, but the benefits of checking for minimum genetic quality standards before replanting are important in vegetable seed production.

Vegetable seed crops must be planted so that sufficient vegetative development occurs to support optimum fruit and seed development. In addition, the increasing emphasis on hybrid seed production necessitates planting of male and female lines at times that assure successful pollination, a process known as nicking. This may require that genetically different male and female lines be planted at differing times so that synchronous flowering occurs. Most hybrid broccoli is transplanted and staggered plantings are necessary for the male and female lines to ensure proper nicking. Hybrid cauliflower and squash are direct seeded but because of the differences in time required for the male and female lines to flower, they are planted in a relay fashion where the earlier flowering line is planted first followed by the later flowering line.

Vegetable seed production also is unique from most agronomic crops because vegetables such as beets, carrots, and cabbage are biennials and must develop sufficient vegetative growth prior to cool temperature exposure in order that vernalisation successfully induces flower formation the following season. Planting such crops too early causes winter kill or late season pest infestations. Planting too late results in a lack of vernalisation which limits flowering and reduces seed yield. In some instances, vegetable seed producers opt not to start the seed crop by seed. Rather, to ensure optimum vegetative growth for vernalisation, they plant some developed portion of the vegetative structure. This is commonly done with carrot and beet stecklings and onion bulbs which allows for maximum production of plant biomass before vernalisation.

Vegetable seeds also vary greatly in their tolerance to soil temperatures at planting. Pea, radish, and spinach require cool soil temperatures for optimum seedling emergence. Beets, cabbage, carrot, and onion are tolerant of cool soil temperatures although they do better in warmer soils. Squash and melons require warm soil temperatures for optimum seedling emergence. In regions with high summer temperatures, vegetables such as tomato, pepper, eggplant, and cucurbits are planted in the early spring to optimise flowering and seed set in early summer. Later plantings of these seed crops are possible by transplanting young plants previously started in glass houses. This avoids the delayed establishment of plants from direct seedings due to cooler night temperatures and wetter soils encountered in the early spring.

Row spacings and planting densities of vegetable seed crops differ from those for fresh market production. Sufficient space for flower development, air movement to reduce pathogens, unrestricted access to inflorescences by pollinators, mechanical cultivation, and harvest operations are necessary for seed crops. In some cases (e.g., eggplant, pepper, tomato, muskmelon, watermelon), row spacings used for seed production are the same as those for fresh market production. In most cases, row spacings are different for vegetable seed production. In lettuce, the size of the head will be larger than the fresh market head in order that bolting is promoted.

Large-head types are planted 25 to 30 cm apart while small-head and loose-leaf types are planted 15 to 20 cm apart. Similar considerations are encountered for cabbage. Broccoli and cauliflower produce sizable seed stalks requiring rows 40 to 60 cm apart. Planting densities are often different as well in vegetable seed crops from fresh market vegetable crops. Beet seed production is accomplished at higher planting densities than beets harvested for their roots. Carrot, celery, and parsnip seed production is planted at densities much less than for root or stalk production. Higher planting densities of onions grown by the bulb-to-seed method reduces lodging of the large seed stalk scapes.

The method (wind or insect) of pollination affects row spacings and planting densities. Vegetables that are wind pollinated include sweet corn and spinach. Insect pollinated vegetables include most of the cole crops, carrot, and onion. Row spacings and planting densities are also important considerations in the production of vegetable seed hybrids. In order to maximise the amount of hybrid seed produced, the optimum ratio of female to male rows must be determined.

Fertilisation

Vegetable seed crops have similar nutrient requirements as fresh market crops. Where possible, nutrients that stimulate reproductive development can be supplemented with normal fertility regimes. Boron, for example, increases the number of fruiting sites and seed produced per pod in legumes such as pea and bean. Molybdenum also is essential for optimum nitrogen fixation of legume seed crops. Another approach to stimulating reproductive development in vegetable seed crops is to split the application of fertilisers at planting and before flowering.

This avoids luxury consumption of the elements at the time of application, improves crop uptake efficiency, and increases the formation of reproductive structures. Generally, lower nitrogen and phosphorus levels are used in seed production because these encourage continued vegetative growth. Because most vegetable seed crops are grown in rows, fertilisers are typically applied either in bands in the planting row or as a side-dressing beside the row after establishment.

Weed and Pest Control

The management of weeds and pests in vegetable seed production poses additional problems due to their longer growing season not encountered for vegetables grown for market use. This extended growing period necessitates an increase in residual pesticide activity or additional pesticide applications. In some cases, governmental regulations do not permit the use of the same pesticide for a vegetable seed crop that is used for the market use crop.

This is because the approval process is directed at the more economically important value crop. Some pesticides which control insects also detrimentally affect insect pollinators essential for cross pollination or hybrid seed production and cannot be used in a seed production field. Such restrictions on the use of pesticides in vegetable seed production imply that the safe use of pesticides will be more difficult in the future and greater reliance on crop rotations and biological control will be necessary.

Insects and the timing of their infestations affect vegetable seed production. The Lygus bug prefers legume crops and safflower as its primary host. However, once these crops are harvested, it commonly flies to adjacent carrot seed fields and rapidly creates epidemic infestations. There, it feeds on developing seed embryos and subsequently reduces seed yield and quality. Cabbage looper is a significant pest of brassicas during vegetative development but does not affect seed stalk and pod development. Thus, control of this pest is important during vegetative growth if maximum yield and seed quality are attained. Aphids affect the harvest and subsequent conditioning of brassica seed crops by the development of "honeydew" which adheres to the seed pods and makes separation difficult.

Vegetable seed crops are susceptible to the same diseases as fresh market crops. However, the need to complete their entire life cycle for seed production means that other diseases may also be encountered and require control. Many significant diseases of vegetable seed crops can be controlled by growing the crop in differing geographic locations. For example, the low relative humidity and high temperatures of the southwestern United States reduce the incidence of the seedborne pathogen Cercospora carotae in carrot seed fields commonly found in the high relative humidity, low temperature Pacific Northwest. The incidence of downy mildew is reduced by growing long-day onion seeds in the low humidity areas of the desert southwestern United States.

Diseases in vegetable seed fields can also be minimised by separating the seed crop from disease pests either physically or over time. Physical separation occurs by producing the seed crop in an isolated region from fresh market crops or from the disease causing organism. Celery seed fields are isolated from celery stalk production fields to reduce the incidence of late blight caused by Septoria. Lettuce seed fields are isolated from leaf and head production fields to reduce the incidence of seed-borne lettuce mosaic virus. Isolation from the disease causing organism is achieved by maintaining weed-free irrigation canal banks and roadsides which serve as alternate hosts or growing the seed crop where the disease causing organism does not exist. To meet phytosanitary requirements for export, onion seed crops must be grown in regions where certain disease and nematodes are certified to be absent.

The lettuce mosaic virus is another serious disease that results in unmarketable lettuce heads. To control this disease, an integrated pest management strategy has been established which focuses on the planting of disease-free seed and control of insects and weeds. Disease-free seed is achieved by planting stock seed in glass houses free of such virus vectors as the green peach aphid (Myzus persicae). Commercial seed lots are certified to have not more than one infected seed in 1,000 seeds that have been screened. Other effective measures include control of common weeds that serve as alternate hosts to insect vectors and frequent roguing of lettuce seed fields, to remove any plants suspected of carrying the virus.

Some disease causing organisms have relatively short life spans. In these cases, separation of the seed crop from the organism over time reduces disease problems. Alternating cabbage and other cruciferous seed crops in a three year rotation with non-cruciferous crops decreases the incidence of club root caused by Plasmodiophora brassicae. Nematodes can similarly be controlled in cucurbit and solanaceous seed crops by three year rotations with crops not from these families.

In other cases, diseases can be controlled by storing the seed for a period during which the disease causing organism dies. An example is soft rot (Erwinia carotovora) on celery seeds which dies after two years of seed storage. Hot water baths or chemical treatments also control diseases. Hot water treatment is used on a variety of vegetable seeds to control a number of seedborne bacteria and fungi. Many disease causing organisms are carried on the seed coat and most of these are controlled by various topical seed treatments consisting of fungicides, insecticides, and bactericides.

Harvesting, Threshing, and Drying

Vegetable seed crops are divided into two categories based on the type of fruit harvested. Dry-seeded fruits include the brassicas, legumes, and onion and wet-seeded fruits include the cucurbits, melons, and tomatoes. The methods of harvest and extraction of seeds vary dependent on the type of fruit harvested.

Of great concern in dry-seeded fruit vegetable seed production is the potential for loss of yield due to seed shattering when harvesting occurs after physiological maturity. To minimise seed loss, two approaches are employed. Since many dry-seeded vegetables achieve physiological maturity before harvest maturity, they are cut while the plant is still green and placed into rows where the wind further dries the plant and its reproductive structure. This permits more uniform maturation of all dry-seeded fruits and seed quality is enhanced compared to the more costly option of sequential harvest of mature fruits.

Later, after the seeds have reached harvest maturity, they are extracted from the seed heads either by hand or mechanically using a combine. The second approach is to visually monitor the development of the dry-seeded fruit and harvest the crop before shattering occurs. For example, onion seeds are very susceptible to shattering once the florets have dried and seed germination and quality is reduced if the scape stalks are cut too early.

A common harvest criterion has been to examine umbels for the first mature seeds. In indeterminate crops such as carrot, seeds mature in sequential order with those on the primary umbel maturing first followed by those on secondary and tertiary umbels. Secondary umbels produce about 50% the quantity of pure live seed of the primary umbel. Thus, harvest of this dry-seeded crop is complicated by the umbel maturation sequence. Seeds produced from the first lettuce flowers also tend to be larger and of higher quality than seeds produced from later flowers.

Seed harvest from wet-seeded fruits is often determined by fruit color. Seed quality in pepper as determined by percentage germination and rate of germination is higher in seeds harvested from red fruits compared to those harvested from green fruits. In tomatoes, maximum seed germination is attained when the fruits are at a mature green stage compared to those which are at the red or fully-ripe stages.

After the fruits are determined to be at the correct stage for maximum seed quality, they are gathered and the seed extracted. In some cases, this is done in one mechanical operation where the fruit is picked from the row, conveyed to a crushing mechanism, and the pulp removed from the seed by washing with water in a rotating drum. For high moisture vegetables such as melons and tomatoes, water is periodically drained from the equipment. For low moisture vegetables such as squash and peppers, water is periodically added to the equipment.

In most cases, vegetable seed companies mechanically harvest crops such as tomatoes, peppers, and cucumber in the field, transport the fruit to a central extraction plant, and remove the seeds from the fruit. In tomato, cucumber, and some melons, a mucilaginous layer around the seed must be removed prior to cleaning. This can be done by two methods. The first, and most common, is the retention of seeds in the crushed fruit pulp for a few hours to more than a day so that fermentation occurs. This process is closely monitored because the fermentation creates heat and mechanical injury to the seeds if continued too long. The second approach is to add sodium carbonate or hydrochloric acid to the crushed fruit extract to degrade the mucilaginous layer.

After the fruit is crushed and any mucilaginous layer removed, the seed is separated from the pulp and passed along a water sluice containing gates along its entire length. Mature high quality seeds sink behind the gates and are separated from the pulp by the running water. After the pulp is removed, the sluice gates are opened and the mature seeds captured on a screen. At this stage, the wet seed must be promptly dried to avoid heating and loss of seed quality with increasing imbibition. To accomplish this, the wet seeds are collected and placed into a spin dryer to remove excess water and the moist seeds are evenly spread on screens and exposed to large volumes of heated air. After drying is completed, the seeds are handled and conditioned in the same way as employed for dry-seeded vegetable crops.

Cleaning and Storage

The physical quality of the seed lot following harvesting, threshing, and drying must be evaluated to determine whether precleaning is necessary. If the seed lot possesses large amounts of inert material such as stones or plant debris, it should be passed over a scalping machine. If the seed possesses undesired appendages which cause the seeds to clump together such as carrot, then it must be passed through a debearder to ensure free-flow of individual seed units. Impact debearders are also used in multigerm beet seeds to split the fruit into single seed units.

After precleaning, vegetable seed crops are conditioned in the same way as other seed crops using the same principles and equipment. Some additional improvements can be made. For example, greater seed density improves the seed quality of lettuce, tomato, and onion crops. This can be accomplished by placing the seed mass into a solution of specific osmotic concentration. The more dense seed fractions are retained and the more buoyant seeds removed. After cleaning is completed, vegetable seeds are stored using the principles of successful seed storage.

It is not possible to exhaustively study the range of seed production schemes utilised in the production of vegetable seeds. Rather, six of the most important vegetables (onion, tomato, cabbage, lettuce, cucumber, carrot, and sugar beet) are considered. These were selected not only because of their economic value but also because they represent differing botanical families, generally differ in their vegetative/reproductive structure harvest, and illustrate differing seed production practices. Similarities exist in the seed production of related species and many of the principles detailed here can be extended to these crops as well.

REFERENCES

Bassett, M. J., *Breeding Vegetable Crops*, Avi Publishing, Westport, 1986.

George, R. A. T., *Vegetable Seed Production*, Longman Press, Essex, 1985.

2

Assessing Seed Vigour

The development of a seed encompasses a series of important ontogenetic stages from fertilisation, to accumulation of nutrients, to seed dry down, to dormancy. Each of these stages represents a change in morphological and physiological development that can alter seed performance potential. The point at which the seed achieves its maximum dry weight is called physiological maturity. At this point, it has its greatest potential for maximum germination and vigour.

However, since seeds generally achieve physiological maturity at high moisture levels unsafe for storage, seed is typically not harvested until it attains harvest maturity, which is low enough for safe storage, but high enough to minimise mechanical injury. Between physiological maturity and harvest maturity, the seed is essentially stored on the plant where it may be exposed to severe environmental conditions that adversely affect seed quality.

SEED VIGOUR TESTS

The challenge of vigour testing has been to identify one or more quantifiable parameters that are common to seed deterioration. Although not all changes that occur during seed deterioration are understood, we can speculate on the probable sequence of events. A hypothetical model has been developed by Delouche and Baskin that outlines some of the major parameters used in measuring seed vigour.

Because a vigour test is a more sensitive index of seed quality than the standard germination test, any of the events that precede loss of germination could serve as a basis for vigour tests. The earlier the parameter can be measured during the loss of germination, the more sensitive the index of seed vigour. Thus, since the onset of membrane degradation precedes loss of germination, the most sensitive vigour test should be one that monitors membrane integrity.

Considerable experimental evidence supports this contention. Membranes are essential for many metabolic events occurring in the seed, including respiration (cristae in mitochondria), which provides the seed with the energy required for subsequent growth. The endoplasmic reticulum is also a membraneous organelle on which many enzymes are formed as ribonucleic acid is translated. Thus, any impairment of membrane function can decrease the amount of ATP formed as an energy source, as well as retard the synthesis of specific enzymes essential for growth. Subsequent to the loss in respiration and biosynthetic capacity, the rate of germination declines, culminating in a loss of seed lot uniformity.

Other associated events that occur during deterioration are loss in storability and ability to resist disease infection. Deteriorated seeds that are subjected to biological and environmental stresses also exhibit reduced field emergence which may be related to final yields of certain crops. Eventually, these subtle manifestations of loss in seed quality are expressed by an increasing incidence of abnormal seedings - a component of the germination test. The final parameter of seed deterioration (and the one most often employed) is seed germination, underscoring the need for seed vigour tests to supplement routine standard laboratory germination tests.

As a result of the biochemical and physiological changes known to occur during seed deterioration, most vigour tests have focused on measuring one or more of these parameters.

Characteristics of a Seed Vigour Test

A vigour test should possess certain essential characteristics that can make it useful to the seed producer and consumer. These characteristics have been described by McDonald as follows:

— *Inexpensive.* Due to limited budgets for seed testing, it is important that a vigour test be reasonably priced and require a minimum investment in labor, equipment, and supplies.

— *Rapid.* Every seed laboratory has periods of peak activity, thus it is important that the vigour test be conducted rapidly to minimise analyst time and germinator space. Furthermore, seed producers desire a quick turnaround time for samples submitted for vigour tests since such quick information on seed quality can provide them with a competitive marketing advantage.

— *Uncomplicated.* Where possible, vigour test procedures should be simple so that they can be performed in seed laboratories without requiring additional staff with special backgrounds and training.

— *Objective.* For a vigour test to be easily standardised, a quantitative or numerical index of quality that avoids subjective interpretations by analysts should be utilised.

— *Reproducible.* The success of any test depends on its reproducibility. If these results cannot be repeated because of intricate procedures or subjectivity of interpretation, then comparison of results among laboratories becomes meaningless.

— *Correlated with Field Performance.* Most definitions of seed vigour emphasise the relationship between seed vigour and field performance, and many studies have demonstrated this association exists. Consequently, the ultimate value of any vigour test may be its ability to predict field performance.

Types of Seed Vigour Tests

The standard germination test is conducted under optimum conditions for seed germination. Consequently, when field conditions at planting are near optimum, the results usually correlate well with field emergence. However, under suboptimal field conditions, standard germination results usually overestimate field emergence. Therefore, additional tests are needed to better predict seedling emergence under a wide range of field conditions. Many vigour tests have been suggested; however, only a few have attained acceptance by seed analysts and seed testing organisations. This discussion will focus on the two most popular vigour tests: cold test and accelerated aging test.

Cold Test

The cold test is one of the oldest methods of stressing seeds and is most often employed for evaluating seed vigour in corn and soybeans. Seeds are placed in soil or paper towels lined with soil and exposed to cold for a specified period, during which stress from imbibition, temperature, and microorganisms occurs. Following the cold treatment, the seeds are placed under favorable growth conditions and allowed to germinate.

The cold test exposes seeds to 10°C for 7 days in non-sterile field soil at approximately 60-70% of water holding capacity prior to a 4-7 day grow out period in ideal conditions (25°C). The moisture and temperature conditions provided in the cold test simulate the adverse conditions that seeds might encounter in an early spring planting. The ability of seeds to germinate and emerge in cold wet soil is affected by genotype, seed quality (both physical and physiological), seed or soil-borne pathogens and seed treatment.

The cold test attempts to measure the combined effects of all of these factors. It often represents the lowest emergence that would be expected from a seed lot when planted under reasonably satisfactory field conditions, while the germination test represents the highest emergence potential that could be expected. When germination obtained in the cold test is very close to that obtained in the standard germination test, the seed lot would be expected to emerge well over a wide range of soil moisture and temperature conditions. There are two common approaches to conducting a cold test: rolled towel method and tray method.

Rolled Towel Method

Equipment and supplies:

— *Germination paper towels*: Paper towels of the same weight and thickness as used for standard germination, 30 x 60 cm in size.

— *Plastic bucket or container*: To hold rolled towels in an upright position for placement in germination chamber.

— *Plastic bags (2 ml thickness)*: To cover upright towels after placing in container.

— *Two germinators*: Two chambers capable of maintaining 10°C and 25°C (+/- 0.5°C).

— *Heat exchange plate*: To maintain temperature to towels at 10°C.

Tray Method

Equipment and supplies:

— *Food service tray*: Fiberglass, plastic or metal tray (45 x 66 x 2.75 cm).

— *Crepe-cellulose paper*: Sheets of creped cellulose (40 x 61 x 0.5 cm) to fit in base or tray which is available from many paper supply houses.
— *Graduated cylinder (500ml) or automatic watering device*: Used for adding specific amount of water to crepe-cellulose and tray.
— *Food cart*: Closed food service cart with plexiglass back to hold trays during testing.
— *Two walk-in temperature controlled chambers*: Two chambers which allow entry of food service cart and are capable of maintaining 10°C and 25°C (+/-0.5°C).
— *Alternative equipment*: If laboratory does not have walk-in temperature chambers for food cart use, two germinators as described for rolled towels could be used and the trays moved manually from 10°C to 25°C.

Both Methods

Field soil: The soil component is a very important aspect of the test and the selection of an appropriate soil source is critical for the reproducibility of the test. Since the soil is expected to supply disease-causing organisms, it should be from a field site which has supported some vegetation, preferably the crop to be tested. Highly organic soils may exhibit lower germination due to excessive pathogenic activity which may cause erratic test results. If possible, adequate soil for the entire testing season should be obtained from the same source. It should be handled in a consistent manner throughout the testing season and stored under uniform conditions to maintain soil moisture. Prior to use, the soil should be screened through a 5 mm sieve to remove large debris. It may be mixed with sand, vermiculite, peat or applied directly. Since the role of the soil in these tests is to provide an innoculum source, the quantity of the soil needed for each test is relatively small.

Seed samples: If cold tests are to be a useful guide in planting decisions, the seeds should be tested in the same condition as planted. They should not be allowed to dry out prior to testing, and they should have recieved their final fungicide treatment. Seeds at unusually low moisture levels (<8% moisture, fwb) may exhibit poor cold test performance due to imbibitional chilling injury; thus, seed lot moisture should be adjusted (if necessary) to a range of 8-13%.

Procedures of Rolled Towel

This widely used procedure was first described in the 1950s. The towel procedure has gained widespread acceptance for its reliability, minimal space requirements, low volume of soil/sand needed, and generally good correlation with other cold test procedures and with field emergence. The common features of most of the current versions of this test recommend that the towel substrate includes soil or a soil/sand or peat mixture as an inoculum source. Following a low temperature (10°C) stress period for 7 days, the rolled towels are transferred to an optimum temperature (25°C) for germination for 5 additional days.

Methods

— At least 16 hours prior to planting, the germination paper towels should be soaked until each towel has absorbed approximately 35 ml of water. The saturated towels should be equilibrated at 10 C overnight to ensure that the entire towel mass is at 10°C.

— A minimum of four subsamples of 50 seeds each drawn from the pure seed fraction of each seed lot are tested. The seeds of each subsample are placed on a double layer of cold saturated towels in two 25-seed rows, 6 and 12 cm from the upper edge of the top towel. The planted seeds are lightly covered with soil/sand substrate (one part soil/one part sand) ensuring that all seeds are in direct contact with the soil. A top towel is placed over the two lower towels containing seeds and soil and the three towels are loosely rolled. Care should be taken during and after planting to ensure that the towels do not warm above 10°C. This can be done by placing the chilled towels in an insulated cooler over ice or by using a heat exchange plate.

— When ready for transfer to the cold (10°C) chamber, the rolled towels are placed upright in a plastic bucket or container. The containers have wire dividers to keep the towels separated and are covered with a plastic bag to prevent loss of moisture.

— The covered containers are placed in a cold room at 10°C (+/- 0.5°C) in darkness for 7 days.

— After the cold treatment, the containers are transferred to a germination chamber at 25°C for 5 days in darkness.

— Seedlings are evaluated using the same criteria as for the standard germination test. If desirable, seedlings may be classified as strong or weak after 7-10 days as described by Fiala.

Tray

Common features of this test include a food service tray, a base of creped cellulose paper which acts as the water reservoir for the seed and a covering of a mixture of soil/sand. Test conditions are similar to the rolled towel test in that the first 7 days are at 10°C, but the growth period is shorter, 4 days at 25°C. Although the test requires considerably more space than the rolled towel procedure, it has several attractive features including: mechanisation and standardisation of substrate watering, vacuum-head or trip-board planting, uniform distribution of seed onto the substrate allowing ease of interpretation, natural emergence of seedlings through a soil/sand substrate allowing faster evaluation and greater efficiency (less time per sample).

The results of the Tray Method are correlated with field emergence and with other widely used cold test procedures. The primary limitations are the initial expense of the equipment and the relatively large amount of soil and floor space required to conduct the test, which may reduce the number of samples tested.

Procedures

— Prior to planting, the water, trays and carts and sand/soil should be chilled to 10°C.

— One sheet of crepe-cellulose is placed into a standard food service tray and soaked with 1,100 ml of chilled water. The addition of water is most conveniently done with an automatic watering system which ensures uniform water distribution across the tray. Water can also be distributed in the tray with a graduated cylinder or other measuring device if desired.

— Each tray accommodates 4 x 100 seed subsamples. Thus, two seed lots of 2 x 100 seeds each are planted in each tray. The 100 seeds in each subsample may be planted with a vacuum head or a trip planting board directly onto the saturated crepe-cellulose in each one-quarter section of the tray. The seeds are pressed uniformly into the crepe-cellulose using the bottom of an unused tray so that seed contact is constant and uniform distribution is maintained during the addition of the soil/sand substrate.

— An excess of the chilled soil/sand mixture (1 part soil/2 parts sand) is spread over the seeds and crepe-cellulose. A board, notched to provide a level surface 0.4 cm below the lip of the tray, is used to level the soil/sand mixture and remove any excess. After the soil/

sand absorbs water from the wet crepe-cellulose, the resulting seedbed should approach 70% of the water holding capacity.

— After planting, the test trays are placed at 7.5 cm intervals in a closed food service cart with the plexiglass back. The cart is moved into a 10°C (+/- 0.5°C) chamber in darkness for 7 days. The carts are then moved to a 25°C chamber with an alternating light source (8-12 h light per day) facing the plexiglass back side of the cart for 4 days.

Interpretation of results

The cold test does not provide an absolute vigour score, but records germination after 7 days stress under cold temperatures and high soil moisture in the presence of soil microbial activity. Since germination can be influenced by both the soil source and the cold test method used, the interpretation of results should be facilitated by the use of a control seed sample of moderate cold test performance with each seed lot or group of seed lots being tested.

The relationship between the germination of the control sample and that of the samples being tested is probably more important than the actual germination of the test samples. Within a laboratory and/or a seed company, the level of cold test germination can be related to field emergence and seed quality decisions can be made regarding potential seed lot performance. Until cold test methods are standardised, the results cannot be interpreted at a given vigour level among laboratories. The cold test is therefore limited primarily as an "in-house" vigour test within a laboratory and/or seed company.

Scope

The cold test has been used extensively for many years for testing corn and sweet corn seeds. The relationship between cold test results and field emergence has ranged from excellent to variable depending on the method used and the level of stress in the field. When seeds are planted under stressful field conditions, however, there have been many comparisons which show higher correlations between cold test germination and field emergence than standard germination and field emergence. The general principles of the cold test have also been applied successfully to soybean and sorghum seed and could probably be applied to other kinds of seed.

The greatest difficulty with the cold test is the lack of uniformity in field soil. Soils differ in moisture, pH, particle composition, and pathogen levels, all of which contribute to divergent results. Use of vermiculite, a more uniform medium, has been suggested as a possible solution to the variability of soil conditions. However, it is widely believed that a cold test requires field soil to be successful. Regardless of these difficulties, cold test vigour rankings of seed lots tend to remain consistent within laboratories, which support this test as a most useful in-house vigour test.

Accelerated Aging (AA) Test

This test incorporates many of the important traits desired in a vigour test. Initially proposed as a method to evaluate seed storability, the accelerated aging test subjects unimbibed seeds to conditions of high temperature (41 C) and relative humidity (around 100%) for short periods (3 to 4 days). The seeds are then removed from the stress conditions and placed under optimum germination conditions.

Principles

The AA test exposes seeds for short periods to the two environmental variables that cause rapid seed deterioration; high temperature and high humidity. High vigour seed lots will withstand these extreme stress conditions and deteriorate at a slower rate than low vigour seed lots. Although aging temperature and seed moisture have the greatest influence on test results, several other factors which modify the results must be controlled during the test.

When conducting the AA test, seeds are weighed and placed on a screen tray, which is inserted into an inner chamber (plastic box) containing 40-50 ml of water. The inner chamber is placed into an AA (outer) chamber and the seeds aged at a specified high temperature for specified time, dependent upon the species. During the aging period, the seeds absorb moisture from the humid environment within the inner chamber and are stressed by high temperatures as seed moisture increases to a uniform level.

It is important that seeds on the screen tray be kept in approximately a single layer at a uniform distance above the water surface, as the layering of seed will influence water uptake and result in variable seed moisture at the end of the test. Seed size will also influence the final seed moisture and germination following aging. Larger soybean seeds had a lower final seed moisture and lower germination than smaller seeds when the same number were placed in the aging chamber. The final seed moisture can be controlled, however, by basing the sample size on a constant seed weight rather than seed number. Recommended and suggested testing variables have now been identified for a wide range of species.

A uniform temperature in the outer chamber must be maintained and the temperature controls should permit a variation of no more than +/- 0.3°C, since minor changes (0.5°C) in the aging temperature will affect germination results. High relative humidity in the outer chamber is also necessary to prevent water evaporation from the inner chambers. A water-jacketed aging chamber is recommended because it provides precise temperature control, no condensation and a uniform aging environment. The door of the outer chamber must remain closed for the entire duration of the test to maintain uniform and constant temperatures. Opening the door for one minute lowered the temperature of the inner chamber 2-3°C and it took 30 minutes to recover to the desired temperature. The most uniform temperatures of the inner chambers are maintained by allowing some air space (2.5 cm) between inner chambers placed on the same shelf in the outer chamber.

Equipment and Supplies

- *Balance*: Analytical balance capable of weighing to 1.0 mg.
- *Inner aging chamber*: A plastic box (11.0 x 11.0 x 3.5 cm) with a lid into which is placed a plastic or wire tray with a 10.0 x 10.0 x 0.3 cm wire mesh screen (mesh 14 x 18). These trays can be purchased commercially.
- *Seed moisture tins*: Temperatures ranging from 103-130°C as described for each species in the International Rules for Seed Testing.
- *Bottle-top Dispensette (Brinkman)*: Range from 0-100 ml, for dispensing 40 ml water from a standard screw-neck bottle or a 50 ml graduated cylinder.

Water: Deionised or Distilled

Outer aging chamber: Water-jacketed incubator capable of maintaining a constant temperature range from 40-45°C +/- 0.3°C.

Alternative outer aging chambers: This may include incubators that have a heating element immersed in water at the base of the chamber. These chambers frequently do not have precise temperature control and require the use of auxiliary control to maintain uniform temperature. When using these chambers, water may collect inside the top of the outer chamber due to condensation and drip on the lids of the inner chamber. If water accumulates on the lid of the inner chambers, condensation can occur inside the inner chamber (under the lid) and water will drop onto the seed that will raise seed moisture levels during the aging period. This can reduce germination and cause excess mold growth. Precautions must be taken to shield the inner chambers to prevent water droplets from accumulating on the lids during the aging period. Do not use dry incubators or ovens as outer chambers.

Procedure

— The inner chamber AA boxes and screen trays should be thoroughly washed in a 15% sodium hypochlorite (clorox) solution and dried after each use to prevent fungal contamination.

— Place 40 ml (+/- 1 ml) of distilled or deionised water in each inner aging chamber and insert a dry screen tray, being certain not to splash water onto the screen surface.

— Determine the inital seed moisture of the sample to be tested using the method appropriate for the species (ISTA, 1999) and if >14.0% (fresh weight basis), dry the seed to 10 - 14% before testing.

— A minimum of 200 seeds, determined on a weight basis, are placed on the surface of the screen tray (approximately one layer deep). More than one inner chamber should be used to obtain the quantity of seeds needed for larger seeded species. Preferably, seeds to be aged should be untreated. However, if seeds of the crop species are primarily marketed with fungicide treatment, treated seeds may be used.

— The lid is secured (not sealed) on each inner chamber. The inner chambers should be placed on a shelf, transported and placed in the outer aging chamber at the same time. Allow an air space of approximately 2.5 cm between inner chambers on each shelf in the outer chamber to assure temperature uniformity.

— Precisely monitor the temperature of the outer aging chamber and maintain temperature at +/- 0.3 C of desired temperature during the aging period. Record the time when inner chambers are placed in the outer chamber.

— Seed in the inner chambers should be removed from the outer chamber at the exact hour (+/- 15 min) specified and planted for standard germination within one hour after removal.

— The conditions for the standard germination test are those outlined in the AOSA Rules for Testing Seeds or ISTA International Rules for Seed Testing.

— Include a control sample with each AA test. At the conclusion of the aging period, but prior to planting for standard germination, remove a small sample of seed (10-20 seeds) from the inner chamber of the control sample and weigh immediately to test for seed moisture (fresh weight basis) using the oven method.

— When AA tests are initiated for many seed lots on the same day, samples should be grouped at approximately one hour intervals between two outer chambers to allow adequate time for samples to be planted immediately after the aging period.

Interpretation of AA Results

The AA test does not provide an absolute vigour or field emergence score, but simply records germination (percentage normal seedlings) after a period of stress under conditions of high temperature and seed moisture. When the AA results are compared to the standard germination results of the same seed lot prior to aging, the AA germination will be either similar to standard germination (high vigour seed) or less than standard germination (medium to low vigour seed). One use of these results can be to rank seed lots by vigour and decisions can be made regarding the storability or planting of each seed lot.

Relationship with Field Emergence

Early storage studies suggested that AA might have utility as a vigour test to predict field performance. Additional studies have shown that this vigour test functions well in forecasting field emergence and stand establishment in a wide range of crop species. In general, when seeds are planted under stressful field conditions, AA germination provides higher correlations with field emergence than does standard germination.

Predicting Storage Potential

The primary emphasis of early investigations of the AA test was predicting the relative storability of seed of several crop species. It was shown that the changes in germination of high and low quality seed lots after a period of time were consistent with those that occurred in warehouse storage. Subsequent studies have verified that accuracy of this test in predicting the life span of a number of different species under a range of storage conditions.

Limitation of AA

Accelerated aging shows great potential as a vigour test for a wide range of species. Although variability in referee results was shown initially, excellent progress has been made in the refinement of techniques and the evaluation of time, temperature and seed moisture interactions. AOSA and ISTA referee tests with soybean seed report greater uniformity among laboratories and confirm the relationship between AA germination and field emergence. Sufficient evidence has now accumulated to show that repeatable results can be achieved using this test.

Thus, the AOSA and ISTA now consider this test standardised and have recommended it as a vigour test for soybean. Even slight modifications in temperature control, sample size or aging time will cause variation in final seed moisture and/or germination which will limit the acceptance of the vigour test.

The accelerated aging test is rapid, inexpensive, simple and useful for all species; it can be used for individual seed evaluation and requires no additional training for correct evaluation.

STANDARDISATION OF VIGOUR TESTS

For any seed quality test to be useful, it must provide reproducible tests. To evaluate reproducibility of various vigour test results, both the AOSA and ISTA have conducted extensive referees. Generally, results of these referees have shown that seed testing laboratories can reproduce their own results on the same sample of seed within acceptable confidence limits. However, when the same tests are conducted by different laboratories, the amount of variability can be unacceptable.

Many possibilities exist to explain the lack of standardisation among laboratories. For example, most vigour tests require a degree of subjectivity. The seedling vigour classification and tetrazolium vigour tests possibly involve the most subjective interpretations because seeds and seedlings or both are separated into categories based on characteristics that are difficult to describe precisely. Even the cold test and accelerated aging test require the classification of seedlings as normal or abnormal.

Variations in temperature, moisture, and other environmental conditions are more critical for tests in which rate of growth or rate of a biochemical process is measured than for those such as the standard germination test which measures the completion of a process.

The cold test would appear to be difficult to standardise because it exposes seeds to soil microorganisms to measure their ability to resist attack under cold, wet conditions. Thus, the standardisation of the cold test procedure in its present state still requires standardisation of the substrate (soil) microflora, a task that would be difficult, if not impossible. The use of sterile media inoculated with specific microorganisms has been suggested, an approach that is probably too simplistic, since it is difficult to culture microorganisms and maintain a constant level of pathogenicity. Also, pathogens behave differently in soil, where there is a complex population of microorganisms, from how they behave in the absence of such an interacting population.

Many species (e.g., legumes) have seeds that are impermeable or only slowly permeable to water. This can affect the percentage of seeds that germinate as well as the length of seedling growth in stress tests and can cause a biased index of vigour in a seedling growth rate test. It can also affect the leaching of electrolytes from seeds in a conductivity test. Dormancy can also confound the interpretation of vigour test results. For example, dormancy can cause lettuce and cereal seeds to germinate poorly in tests conducted at high temperatures. If dormancy is suspected of affecting vigour results, the seed should be checked by a standard warm germination test or a tetrasolium test.

Control Samples in Vigour Testing

An essential component of any laboratory testing program is the use of control or check samples. The primary purpose of control samples is to provide internal quality control to enable seed technologists to detect variations in test procedures. Seed vigour tests require precise time, temperature, and moisture that must be monitored and controlled during testing. Even small variations in these components can produce dramatic fluctuations in test results. The selection, maintenance, and use of control samples for each species tested is essential for laboratories engaged in vigour testing.

Control Seed Lot Selection

Seed lots of each crop species tested should be selected annually by screening several samples for both germination and vigour. The sample selected should be relatively free of physical injury and disease, high in germination and moderate in vigour. Seed lots with extremely high vigour will not allow detection of minor differences in methodology and, therefore, would not make good control samples. For example, an acceptable control sample for corn or soybean may have a germination exceeding 95 or 90%, respectively, and vigour level of 70 to 80% and 80 to 90% for the accelerated aging test for soybeans and the cold test for corn, respectively.

Seed Storage and Maintenance

Store enough seed of the control sample to provide an adequate supply for an entire year or testing season. Store the control samples in conditions that will preserve both germination and vigour. Prepackaging the control sample into sub-sumples needed each day may be advantageous, removing from storage the number of samples needed for a week of testing. Two alternatives should be considered for storage and use of control (in order of preference):

— Divide the control seed lot into many subsamples of ~500 seeds. Each subsample should be packaged in moisture-proof foil packets (heat sealed) or stored in a moisture-proof container at low temperatures (0 to -20°C) prior to testing. Control subsamples must be allowed to equilibrate to room temperature (4 to 6 hours) prior to use.

— If it is not possible to handle the small control samples described above, subdivide the control sample into larger subsamples. Place these samples into moisture-proof containers and store at low temperatures (0 to 10°C) prior to testing. Remove small subsamples from the larger samples as needed for testing, place in sealed containers and allow to equilibrate to room temperature before use.

Seed Moisture of Control Samples

The initial seed moisture of the control seed lot must be an acceptable seed moisture for the seed being tested. The moisture of the seed lot must be measured periodically to be certain that the seed has not decreased or increased in moisture during storage. Adjust the moisture to the initial level if necesary, as extremely dry seed can cause imbibitional injury and inaccurate vigour readings during testing.

Use of Control Samples

Control samples should be included as routine samples in every vigour test run and should not be identified as control or check seed until the test is completed. At the conclusion of the test run, the seed technologist or laboratory manager should determine if the vigour test results of the check are within the set parameters. If the sample falls outside the parameters, the problem should be identified and corrected and the test repeated. At frequent intervals, the vigour test results of the control samples should be summarised to evaluate uniformity of the samples and testing procedures.

Clearly, vigour testing for all species and regions has not yet achieved the same level of standardisation possessed by the standard germination test. However, both seed testing organisations (AOSA and ISTA) and the seed trade are focusing on improving the reproducibility of vigour tests. Considering the important role that vigour tests have in plant breeding, seed production, quality control, and marketing programs, standardisation is badly needed. Although future research and testing are required before vigour testing becomes a routine phase of seed testing, the promise of vigour testing in the future is bright.

USE OF SEED VIGOUR TESTS

Many farmers know from experience that seed lots with equal germination levels may emerge from the soil quite differently, resulting in erratic field stands, replanting (on some occasions), or both. These same farmers recognise that this problem is usually more severe when adverse environmental conditions occur at or immediately after planting. Thus, if farmers know that they will be planting seed in adverse soil conditions or can predict an adverse field environment following planting, high vigour seeds should provide higher field emergence than low vigour seeds.

If this prevents replanting, and the associated delayed maturity or yield reductions due to poor stands, additional money spent on high vigour seeds may be a worthwhile investment. It cannot be assumed, however, that high vigour seeds will produce excellent emergence and stands in any soil environment. It should improve the chances for satisfactory emergence, but it will not guarantee it. Conditions prevailing after planting may be so severe that even the most vigourous seeds cannot perform satisfactorily.

REFERENCES

Byrum, J. R. and L. O. Copeland, "Variability in vigour testing of maize (Zea mays L.) seed", *Seed Science & Technolology*, 1995.

ISTA, "Cold Test", *Handbook of Vigour Test Methods*, International Seed Testing Association, Zurich, Switzerland, 1987.

____________, "International Rules for Seed Testing", Seed Science & Technology, (Supplement), 1999.

McDonald, M. B., "Assessment of seed quality", *HortScience*, 1980.

3

Seed Testing

Seed testing is the cornerstone of all other seed technologies. It is the means by which we measure the viability and all the physical factors that regulate the use and maintenance of seeds. Everything that is done with seeds should have some test information to guide the work and ensure high quality. Seed tests tell if a crop of seeds is worth collecting, if handling procedures are correct, and how many potential seedlings are available for regeneration.

The earliest form of seed analysis, the cut test, is still often used today. Before seeds are collected in the field, some seeds are cut open with a knife or razor blade to see if their internal tissues are fully developed and undamaged. This analysis is made more accurate in some cases by the use of a hand lens. It is also used for simple analysis during extraction and cleaning, or after germination to determine if the ungerminated seeds have deteriorated or remained dormant.

Although the cut test is often very good at producing some information quickly, it is limited in the amount of information it can supply and it lacks accuracy compared to more sophisticated procedures. Therefore, it should never be taken as a substitute for a formal laboratory analysis.

SAMPLING

Formal seed analysis begins with the sampling of the seedlot. Sampling can be done with the hand or with a seed probe, also known as a trier. If a probe is used, it must be long enough to reach to the farthest edge of the container. A probe has gates that prevent seeds from entering until the probe is inserted the full dimension of the container. The probe should be inserted into the seed container with these gates closed. Otherwise, seeds from the upper layers will fill the probe as it is inserted and the bottom layers will not be sampled.

Once the tip reaches the bottom or far side of the container, the gates should be opened and the probe gently turned back and forth to help the seeds fall in. Then the gates should be closed gently, not forced, so that any seeds that are caught in the opening and are preventing the gates from closing fully are not crushed. After the probe has been withdrawn from the seed container, it should be held horizontally, with the gates facing upward. Then the gates should be opened gently and the probe shaken gently back and forth, so that seeds caught in the gates will slip down into the probe and the gates can be safely closed. Finally, the probe should be emptied by pouring the seeds out the top of the probe and into a second container. This sample is the first primary sample.

If there is only 1 container, primary samples should be taken until there are 5 primary samples. When more than 1 container holds the seedlot, at least some of the other containers must be sampled. When there are between 1 and 5 containers, all containers should be sampled, at least 1 probe from each container. When there are more than 5 containers, 5 of the containers plus 10% of the remaining ones should be sampled. It is never necessary to sample more than 30 containers. (It would be rare that a forest seedlot would need 30 containers or more, or possibly even 20.) All of the primary samples are then placed together to make up the composite sample. Sampling by hand is sometimes necessary when the seeds will not flow into the probe because of their size, shape, or surface texture.

Sampling by hand can be done by inserting the open hand into the seeds, closing it once the point of sampling is reached, and then withdrawing it closed. The seeds are then placed in a second container to form the composite sample, just as in sampling with the probe. At least 5 handfuls must be taken, and all levels must be sampled. When the hand cannot be inserted into the seedlot, the seeds can be poured from one container into a second, stopping at a minimum of 5 evenly spaced intervals and removing a handful of the seeds for the composite sample.

The composite sample, whether taken with a probe or by hand, is usually too large to submit to a seed laboratory for analysis. The composite sample is, therefore, mixed and divided to obtain a submitted sample. This procedure is very important and must be done correctly for the results to be accurate.

The composite sample can be mixed either mechanically or by hand with rulers. Hand-mixing the composite sample is done by pouring the seeds into a cone on a flat, clean surface. An open file folder makes a good work surface that can be picked up to return the seeds to a container. With one ruler held stationary against the seeds, the second ruler is used to pull the outer edge of the pile up to the top of the pile, allowing the seeds to roll down the sides and over the top of the stationary ruler. The full pile is thoroughly turned over and all layers mixed together. This procedure should be repeated for 1 full minute. Then the pile should be divided by cutting the cone in half and then into quarters. The quarter is then weighed to see if it is enough for the sample. If not, then another quarter, an eighth, or a smaller fraction is taken until the minimum weight is obtained.

Hand mixing can be replaced by mixing with either a soil divider or a gamet divider. These devices can save a substantial amount of time and also, by reducing the tedious nature of the work, increase the likelihood of doing a quality job. The seedlot needs to be poured through the divider 3 times. When the gamet divider is used, the motor must not be turned on until the seeds have been poured completely into the hopper. Once the seeds are cleared out of the machine, the motor must be turned off before the seeds are poured back into the hopper for the next pass. The seedlot is then divided in half, then quarters, eighths, and so forth, to obtain the correct weight for the submitted sample, just as in the hand mixing and dividing.

The size of the submitted sample for some species is stated in the Rules for Testing Seeds and is twice as large as the minimum amount for the purity test. This amount is different for each species and the rules need to be consulted to be sure the correct amount is submitted for purity tests that are to be done according to the rules. A smaller sample of seeds can be submitted, but the test will not be according to the rules and the accuracy cannot be assured to the same degree as a test that is done according to the rules. If a species is not listed in the rules, an amount that contains 2,500 seeds should be taken. This amount can be estimated by counting out 100 seeds and multiplying their weight by 25.

Samples can be as small as 600 seeds when only germination is tested. It is important to work quickly when drawing the sample, if the submitted sample is to be tested for moisture content. This will prevent the gain or loss of moisture from the air. Once obtained, the submitted sample should be put in a moisture-proof container to maintain its true moisture content until it is sampled and tested at the laboratory. Plastic bottles with tight-fitting lids or tightly closed plastic bags of at least 0.1 mm (4 mil) thickness are adequate. Metal containers can be used but are harder to find. Glass containers should not be used; they easily break in transport, allowing the samples to be exposed to the air or, worse, mixed together.

SAMPLE IDENTIFICATION

Assignment of a test number is the first step in handling every seedlot that is received in the laboratory. This number allows for the orderly tracking of the test sample among the other samples in the laboratory. A typical test number indicates the test year and an accession number. For example, the 300th test conducted in 1996 would have a number such as 6-300.

MOISTURE TESTS

Moisture tests must be the first tests conducted on samples when they arrive at the seed laboratory. Once a sample container is opened and work begun, the seeds will likely gain or lose moisture in exchange with the ambient air. The standardised laboratory test for moisture content is the oven method. This procedure was determined, after many years of research, to be a best estimate of moisture for general testing work. This test is made on 2 subsamples containing 3 to 5 g of whole seeds. These 2 samples are placed in containers with lids and weighed to determine the wet weight. Then they are placed in a forced-draft drying oven for 16 to 18 hours at 105 ± 2 °C. The lids are removed during drying but are also placed in the oven.

The samples are then placed in a desiccator to cool for about 20 minutes before being weighed a second time to determine their dry weight. The lids are placed on the cans while cooling and weighing. The loss of weight represents the weight of water in the undried sample. This water weight is divided by the wet weight to obtain the percentage moisture content on a wet-weight basis. The percentage moisture is expressed on a wet-weight basis because this value most accurately represents how much of the seedlot is water. Therefore, the buyer knows how many seeds are purchased and how much water is purchased.

For example, when the price per weight is the same, a pound of seed at 7% moisture content is a better value than a pound of seed at 9% moisture content. For example, a 100-kg seedlot (or a 100-lb seedlot) at 7% moisture contains 93 kg of seeds, whereas the lot at 9% moisture contains 91 kg of totally dry seeds, 2 kg less. Some larger seeds and seeds with impermeable seedcoats need to be cut to make an accurate test. If the seed is not cut open, the moisture is not freely released, and the moisture content is underestimated.

The oven method is not a direct measure of the content of water. It measures weight loss that is assumed to be due to the loss of water. A basic analytical procedure is required to verify the temperatures and length of drying. The currently accepted procedure is the Karl Fisher procedure. The moisture committee of the ISTA uses this procedure in its work to standardise and validate the oven procedures. Another widely used method to measure seed moisture is the electronic moisture meter.

Although there are numerous brands of electronic moisture meters on the market, not all of them will work for tree and shrub seeds, and those that do will not have calibrations for tree seeds. Therefore, conversion charts must be developed for them by testing samples with high to low moisture contents with both the meter and the oven. A linear regression between the oven and meter readings is calculated, and the conversion chart predicted from this regression. These meters provide quick results, are nondestructive to the seed, and are usually accurate to within ± 1% of the moisture estimated by the oven method.

PURITY, NOXIOUS WEED CONTENT, AND SEED WEIGHT TESTS

Purity, noxious weed content, and seed weight tests are sometimes called physical tests because they do not relate to viability. These tests are described individually as follows.

Purity Analysis

After samples for the moisture-content test are withdrawn, the remainder of the submitted sample should be mixed and divided to obtain the working sample, which contains the minimum weight for conducting a purity analysis. Each species has its own specified minimum weight, which has been determined to contain 2,500 seeds. The mixing and dividing should be done in the same way as described in the sampling section for drawing the submitted sample from the composite sample.

However, at this point it is necessary to be very close to the minimum weight for 2 reasons. First, the analyst does not want to examine more seeds than necessary, and second, the accuracy of the test is evaluated using tolerance tables that were developed using these minimum weights. Using substantially more seeds than the minimum will invalidate the use of these tables.

Purity is determined differently by each of the 2 major testing organisations. The ISTA rules specify a 3-part purity and the AOSA rules specify a 4-part purity. The ISTA purity values report percentages of pure seeds, other seeds, and inert materials. The pure-seed fraction consists of all those seeds that are of the kind specified on the seedlot=s label.

Noxious Weed Examination

The noxious weed exam is a specialised purity examination. It is not a test traditionally associated with forest seeds but may become more common as the commercial exchange of native plants increases. A noxious weed is a highly aggressive competitor or a plant with other highly objectionable characteristics, such as being poisonous. It is so offensive it has been put on a noxious weed list compiled by an individual state or the federal government. A noxious weed exam is made solely to identify the number of noxious weed seeds found in the sample. Nothing else is noted in this exam. The presence of any noxious weed seeds makes it illegal to sell the seeds until the noxious weeds have been removed. The sample size for a noxious weed examination is 25,000 seeds.

Seed Weight Determination

The number of seeds per unit weight (kilogram and gram or pound and ounce) is determined on the pure-seed fraction from the purity test. This test is called the seed weight determination in the ISTA rules. It is made by counting out 8 replicates of 100 seeds and weighing them to the same precision as the weights for the purity test. The coefficient of variation for these 8 values is computed. This coefficient cannot be greater than 6 for chaffy seeds or greater than 4 for all other seeds. Otherwise, an additional 8 replications need to be counted and weighed and combined with the first 8 weights.

All 16 weights are then used to compute the mean. Any weight diverging from the mean by more than 2 standard deviations is discarded; only the remaining weights are used to compute the number of seeds per unit weight. Seeds can be counted by hand, with a counting tray, a shutter box, or a vacuum counter. When seeds are counted by hand, it is usually best to count out the appropriate number of piles of 10, 20, or 50 seeds, in order not to lose one's place.

A counting tray is simply a block of wood or plastic with impressions drilled into it to hold the seeds as they are poured across the plate. The counting tray may or may not be faster than counting by hand, depending on the seeds' size and shape of, which determine how many double- or triple-seeded holes must be thinned by hand. Using the shutter tray is similar to using the counting tray, but the shutter tray is emptied by sliding the bottom tray rather than turning it over. For uniformly sized seeds, using either the counting tray or the shutter tray can speed up counting considerably.

The vacuum counter is probably the fastest and one of the most common ways to count seeds in the laboratory. The vacuum counter is made of an acrylic (or sometimes metal) plate that is drilled with 25, 50, or 100 holes and attached to a hollowed-out second plate. A vacuum line is attached to the head and a shut-off valve controls the application of the vacuum. To use the vacuum counter, the seeds are spread out loosely in a 1-seed-deep layer, the counter is placed on top of the seeds, and the vacuum is then turned on. Moving the counting head back and forth for about the diameter of a seed will bring the seeds into contact with a vacant hole. With the vacuum still on, the seeds can be transferred to a dish for weighing or to a germination container.

Some users of vacuum counters report a tendency for lighter seeds, such as empty or partially filled seeds, to be picked up in preference to heavier seeds. To eliminate this problem, this device must be used according to the procedure described above. Seed weights are sometimes determined with an electronic counter. The ISTA rule calls for counting all pure seeds in the working sample when this is done. No error-check is then made.

A recent internal report made by the Seed Count Committee of the Association of Official Seed Analysts, augmented by the author's personal observations, suggests caution in the use of electronic counters for seed weight determinations. A high potential for error in counts exists. If carefully calibrated, these machines can count quite accurately, but the machines need to be adjusted and used correctly. A thorough evaluation of the degree of desired accuracy and the amount of time required to achieve it needs to be made before deciding to use the electronic counter.

Germination Testing

Germination testing is designed to estimate the maximum number of seeds that will produce a normal seedling and to give results that are as repeatable as possible. Without uniform procedures, there would be no standard on which to base the value of seedlots for commercial transactions and the seed trade would be chaotic and filled with dispute. Germination also tells a grower about a seedlot's potential. A seedlot with 80% germination cannot produce more than 80 seedlings/100 seeds. Therefore, if 100 seedlings are needed, a minimum of 125 seeds must be planted (100/.80 = 125).

The germination test is conducted on the pure-seed fraction from the purity test. Both the AOSA and ISTA prescribe the use of 4 replications of 100 seeds. These replications can either be planted 1 to a container, 2 to a container, or all on 1 tray. Alternatively, the 4 replications can be further divided into smaller replications, but the total number of seeds tested must remain 400 to remain in compliance with the rules.

If fewer than 400 seeds are available, then the number of seeds per replication should be reduced so that an equal number of seeds is present in each of the 4 replications. Using fewer than 100 seeds in a replication is not according to the rules, and the test would thus be unofficial. However, it is better statistically to have 4 replications of 50 seeds each rather than 2 replications of 100 seeds each. The 4 replications are then placed under optimal germination conditions for the period specified in the rules. Germination is the number of normal seedlings produced from 100 pure seeds expressed as a percentage.

A normal seedling has all the essential plant structures necessary for the plant to continue to grow normally under favorable conditions. Seeds can be planted in a number of ways. They can be scattered or placed one at a time with forceps, although more generally a vacuum counter or other type of planting plate is used for speed and to ensure even spacing of the seeds. The vacuum counter is the most expedient technique, because it can handle a variety of seed sizes. Counting devices are described in the seed weight discussion above. Seed should be hand-planted only when counting devices cannot be used in order to save time.

Seeds can be germinated on various media. Sand, sand and perlite mixtures, potting mixtures, soil, and various papers—blue blotters, white blotters, or crepe-cellulose papers—can be used. Testing rules, however, specify what is an acceptable medium for the kinds of seeds tested. Specifying the medium helps assure uniformity in test results. The blotters resist penetration by the roots of the plants, whereas the crepe-cellulose paper allows for root penetration. Blotters offer the advantage of keeping the roots where the analyst can actually see them for evaluation, but if a seedling is very large it will fall over and tangle with other seedlings, making counts difficult.

The media also differ in their water-holding ability. Blotters usually need to be watered several times during the test, whereas crepe-cellulose paper, sand, sand mixtures, potting soils, and soil are absorbent enough to hold all the water the seeds need for up to 3 months, if kept in a moisture-proof container. Watering the medium can be done by hand or by machine. Watering by hand is usually done using a squeeze bottle or a small hose from the tap and requires subjectivity on the analyst's part to estimate that the correct amount of water has been applied.

Too much or too little is harmful, but in most cases there is wide latitude in the amount that will give optimal results. Machines for watering include automatic pipetting machines or small traveling spray booms. Both save a great amount of time if many tests are conducted and, once adjusted, take all the guesswork out of applying the correct amount. These machines should be checked periodically, however, to verify that they are in fact applying the desired amount of water.

Germination tests should be run in cabinets or rooms that meet exacting requirements for temperature and light control in order to make accurate and repeatable estimates. Temperatures should be carefully checked throughout the chamber at the level of the substrate to be sure there are no places that deviate from the desired temperature by more than 1°C. Poor air circulation and hot spots from lights or light ballasts are the most common causes of temperatures that are too high or too low. The temperature at which the germination chamber is set depends on the species being tested. Many species do well at an alternating 20 and 30°C. For this regime, the chamber is held for 16 hours at 20°C and for the remaining 8 hours of the day at 30°C.

Other possibilities are constant temperatures of 15, 20, or 22°C, with light usually supplied for either 8 or 16 hours. When temperatures alternate, the light is provided during the higher temperature to follow a natural cycle of light and temperature. Sources of light need to contain abundant amounts of blue and/or red light but not far red light because far red light is known to inhibit germination. Cool white flourescent lamps are most commonly used.

The temperature/light regime used for a germination test is determined by experiments that germinate the same seedlot at different temperature/light combinations. The combination that supports the highest percentage of germination in the most reasonable time is the one that is then adopted in the rules for testing. Dormancy is the condition of a seed that prevents it from germinating when it is placed in conditions that are favorable for germination.

Dormancy must be overcome in order to conduct the germination test, just as when trying to grow seedlings. Prechilling (also called prechill and traditionally called stratification) is the procedure most used for breaking dormancy in forest seeds. The seeds are held in moist conditions at temperatures between 0 and 3°C. Prechilling can be accomplished in 1 of 3 basic ways. In the first, the seeds can be planted on moistened germination medium in sealed containers and then put in the cold. In the second, the seeds can be placed in a moist medium, placed in the cold, and then at the end of the prechilling period, planted on the germination medium.

In the third method (similar to the second), the seeds are soaked for 16 to 48 hours in water to become fully imbibed, placed in a moisture-proof container, held in the cold for the specified period, and then planted on the germination medium. This last procedure is sometimes called naked stratification, because no moisture-holding medium is used. How long seeds are held in prechill varies widely by species and genetic source of the seedlot. The period can range from 10 days to many months. For some species, a warm period preceding the cold period is required. This is called warm stratification or warm incubation. Western white pine and European ash have been reported as requiring this warm-cold stratification.

A species that does not require prechilling is called nondormant. If 10 to 14 days of prechilling are needed, the dormancy would be considered light. If 30 to 60 days of prechilling are required to break the dormancy, it would be considered moderate. More than 60 days of prechilling classifies the seedlot as highly or strongly dormant. The degree of dormancy varies within the seedlot of even lightly dormant species; some seeds germinate without prechilling, whereas other seeds in the same lot will not germinate until they are prechilled. However, the term variable dormancy is usually reserved for seedlots in which some seeds germinate during prechilling, whereas other seeds in the same lot will not germinate even after being placed in favorable germination conditions.

Species that fit the deep and variable dormancy category are Rocky Mountain juniper (*Juniperius scopulorum Sarg.*) and basswood (*Tilia americana L.*). Because of the above-mentioned variation in dormancy, seedlots will often be tested with and without prechilling or with varying lengths of prechilling. Such tests are referred to as paired or double tests; usually only 2 tests are done. More tests, of course, can be and are done with some seedlots. This type of testing can determine the presence of dormancy, the strength of dormancy, or a weakness in the seeds.

When the seedlot has the same germination with and without prechilling, it is said to be nondormant. When the germination is increased with prechilling, the seedlot is classified as dormant; the longer the prechilling period needed, the stronger the dormancy is said to be. A decrease in germination with prechilling is an indication of weakness in the seeds. This last condition is similar to the situation of the type of vigour test known as the cold test, which is described in the following section.

Prechilling is not the only treatment to break dormancy. Light is useful to break dormancy and can reduce the need for prechilling. Birches (*Betula L.*) and loblolly pine (*Pinus taeda L.*) are prime examples where light helps break dormancy. Seedcoat dormancy is treated by scarifying the seedcoat with either acid, bleach or mechanical means. Chemical stimulates such as gibberellins or potassium nitrate have been little used with forest tree seeds.

Vigour Testing

Sometimes standardised laboratory germination procedures are criticised as not predicting field performance very well. These critics suggest using a variety of test conditions to find an optimum for each seedlot. The problem in predicting field germination is that it is impossible to predict the weather with the necessary precision. Vigour testing is one possible solution. The vigour test does not predict performance for a particular set of fluctuations; rather, it predicts the general ability of a seedlot to germinate normally over a range of adverse conditions.

Its purpose is to differentiate seedlots, with essentially equal germination, according to their ability to germinate well in spite of adversity. As seeds age and begin to weaken and die, vigour declines before germination test results decline. Like germination tests, vigour tests are conducted under standardised conditions in order for the results to be repeatable and useful in the field. A vigour test cannot make up for poor practices that unnecessarily increase environmental variation in the field; such poor practices can be major sources of disparity between laboratory and field germination.

The most common vigour tests in agriculture are the cold test, the accelerated aging test, the conductivity test, and the tetrazolium test. These 4 tests have not been used very much for forestry. In addition to these tests, speed of germination as expressed in a number of formulas has been put forward for use in forestry as a vigour test. Despite the potential benefit for tree seed nurseries, the science and technology are not advanced enough to permit the practical application of vigour testing with forest species.

The cold test is done by planting the seed in damp soil and then holding the germination tray at 10°C for 7 days. This test mimics the cool damp conditions of soil in early spring. At the end of the period, the germination trays are transferred to the appropriate temperature for germination. The higher the percentage of germination, the more vigourous the seed is said to be. High-germinating lots have the needed strength or vigour to pass the period of stress and still have energy for high germination when conditions are favorable. This is analogous to the case of the tree seedlot that drops in germination following prechilling. Those that drop in germination after prechilling are weak.

The accelerated aging test is conducted with the stress of high temperature and moisture. The given weight of seeds is placed in a small box with a screen tray that suspends the seeds over a reservoir of water. These boxes are then placed in an aging chamber at 40 to 43°C for 72 to 288 hours, depending on the species. Whichever temperature is chosen in this range, the variation must be virtually nil to ensure repeatability of the results. The water-jacketed incubator has been determined by organised tests among laboratories to be the best device to give this necessary strong control over the test conditions. At the end of the period the seeds are planted and tested for germination under the standard conditions.

The electrical conductivity test has been widely tested in agriculture but has not been adopted as routine practice except in a few specialised areas. In this procedure, seeds are soaked individually or in bulk. Deteriorated or dead seeds leak electrolytes more readily than high-vigour seeds. This greater leakage causes the water to have a higher conductivity, which can be measured with a conductivity meter. Bonner and Agmata-Paliwal reported on the use of conductivity for tree seeds and found that results have poor repeatability for precise estimates but possibly would work for general estimates of classes as poor, low, intermediate, or high viability.

Several statistics have been put forward to use speed of germination as an indicator of vigour. The faster a seedlot completes germination or reaches its peak the more vigourous it is said to be. The simplest indicator is days to 90% of total. For example, if the final germination is 88%, the indicator would be how many days it takes to reach 79% germination. A lot that reaches 79% in 12 days would be more vigourous than one that takes 16 days. To use this statistic, counts must be made quite frequently, even daily, or the data must be interpolated to determine the number of days to the specified germination.

Czabator's factor, developed for use with southern pines, combines the maximum daily average germination, called the peak value, and the average daily germination at the end of the test to form one statistic called the germination value. Germination is counted frequently, at least every third day, and the cumulative germination on each day is divided by the number of days that the test has been run in order to compute the mean daily germination for that day. For example, if on day 22 the cumulative germination is 88, the mean daily germination is 4.

This mean daily germination increases with each day of germination until the period of maximum germination has ended and then decreases. The largest value of the mean daily germination is called the peak value. Germination value is computed by multiplying the peak value by the mean daily germination. Lots that have higher germination values are generally considered more vigourous.

Another characteristic of more vigourous lots is that they store for longer periods of time without loss of germination. Therefore, if 2 seedlots have equal germinations, the one with the lower vigour might be considered for first use, because the germination of this lot will likely decrease faster than a lot of higher vigour. This approach would give the greatest potential number of seedlings. Lower vigour seedlots will loose viability even under ideal conditions in the freezer. Tetrazolium staining has been tried also as a vigour test. Because of the highly subjective nature of this test and the great amount of experience it requires to administer, it has never been widely used as a vigour test and never successfully with forest plants. As stated in the next section, tetrazolium can be used to successfully estimate viability for very dormant species or for other hard-to-germinate species.

The problem with vigour analysis is that it has proven to be difficult to standardise. This difficulty is apparently because the test conditions are so exacting. In tests involving germination, the temperature must be very tightly controlled. Just a degree or two difference in temperature can change the speed of germination, effecting the value of those statistics that rely on germination speed. The accelerated aging test was difficult to standardise until chambers were developed that had virtually no variation in temperature. Difficulty in standardizing laboratory tests and the lack of clear and consistent interpretation to the field has prevented the operational use of vigour testing.

Chemical Staining for Viability

The tetrazolium staining procedure mentioned in the vigour section is useful in estimating the viability of dormant seeds, especially very dormant seed. This test involves soaking the seed first in water to fully imbibe the seed and soften it for cutting. A moistened seed will take up the stain more rapidly. A variety of methods are used to open seeds. It is extremely important that no damage occur to the embryonic axis when a seed is cut. The embryonic axis is the radical and the plumule. The meristematic regions are here, and their condition needs to remain unaltered until they are carefully examined. These are the areas where the embryo must grow in order to produce a normal seedling.

Usually forceps and sharp single-edged razor blades are used to cut open the seeds. For seeds with harder or stony seedcoats, a variety of vises, hammers, and clippers are used to cut through or remove the seedcoat. The solution that is used to make a tetrazolium (TZ) test is colorless. It is made by dissolving 2,3,5-triphenol tetrazolium chloride in a phosphate buffer at pH 7.4, which is the optimum pH for the TZ reaction. The buffer is necessary to compensate for any pH imbalance in the TZ salt, the water, or possibly the seeds. The colorless solution is taken up by the prepared seeds and then reacts with respiratory enzymes (that is, dehydrogenases) to form an insoluble light pink (magenta) precipitant.

Tissues that are alive and respiring will stain, and those that are not alive will not. For a detailed discussion of this procedure, refer to the AOSA Handbook on Tetrazolium Testing. The TZ test can be completed in 4 to 48 hours, depending on the amount of preparation time required and the rate of staining. Tetrazolium staining has proven useful with many species that have deep dormancy, including tuliptree (Liriodendron tulipifera L.), baldcypress (Taxodium distichum (L.) Rich.), Rocky Mountain juniper, and sumac (Rhus spp. L.) species. However, for a few species with very deep dormancy, there will be no staining unless the seeds are prechilled.

Excised Embryo Testing

The excised embryo test is done on the embryo after it is removed from the seed. In this germination test, the embryo has been freed from the restriction of the seedcoat and nutritive tissue. Therefore, a germination that would take many months and be incomplete can be complete in 10 to 14 days. The ashes (*Fraxinus L.*), maples (*Acer L.*), and cherries and plums (*Prunus L.*) are some of the genera that are tested by embryo excision. Because the embryos are very vulnerable to infection once excised, the test must be done under strictly clean (axenic) conditions.

The work surface and all tools, hands, and germination dishes should be washed carefully, perhaps with absolute ethanol. If good embryos easily mold, then cleaning procedures must be reviewed for effectiveness and the work area examined for sources of microbial contamination. Generally, however, sterilisation procedures such as autoclaving are not required. The procedures for excising the embryos are similar to those used in preparing seeds for tetrazolium. Greater care is needed, however, because the embryo must be removed intact without any significant injury or broken apart.

The main advantage of this method over the TZ test is that the evaluation is less subjective; the growing embryo is actually observed in most cases. Therefore, a direct reading on the growth potential of the seedlot can be obtained. A complete excision is not always required. Russian-olive and western white pine respond well to only partial removal of the seedcoat. The white pine requires prechilling before the seedcoat is cut open.

X-radiography

X-radiography is very useful in forest seed analysis. It provides a very rapid and accurate analysis of the internal structure of seeds, identifying empty, insect-damaged, or poorly developed seeds. It is an immense help in judging maturity, determining how many bad seeds should be remove, and detecting any mechanical injury. It is more accurate than cutting tests in many cases, requires much less time, and is nondestructive . X-radiography was first applied to tree seeds by Simak in Sweden.

The use of contrast agents has improved the ability of the x-ray test to discriminate between viable and nonviable seeds with some species. A contrast agent enters damaged areas of the seed differentially from nondamaged areas, making the damaged areas more radiopaque. They will then appear as bright areas on the radiograph. Aqueous solutions of heavy salts such as iodine or barium chloride and vaporous agents such as chloroform have been used as contrast agents.

Radiographs can be made on Polaroid7 film, x-ray paper, or x-ray film. Polaroid film is useful if no darkroom is available,because the film is developed in the light, just like a Polaroid photograph. The disadvantage of Polaroid is the high cost, short shelf life, and lack of detail. Xray paper is fast to use but does require a simple darkroom. It is less expensive than Polaroid, has a shelf life of several years in cold storage (3 °C), and much better resolution. The best resolution is obtained with x-ray film. X-ray film, however, is more expensive and slower to develop (over 45 minutes) and a light table is required to see the images.

Usually the film or paper is placed in a paper or vinyl cassette so that it may be handled in the light. This cassette can result in some loss of clarity of the image, especially with small seeds; using the film in the dark where the seeds can be laid directly on the emulsion gives a noticeably superior image. Seed work is usually done with x-rays in the range of 10 to 30 kvp (that is, kilovolt potential), which is is the amount of penetrating power the x-rays have. The exact kilovolt potential depends on the equipment and the seeds in question. Trial and error is necessary to find the best combination. Too high a kilovolt potential and the seeds will not be visible or will appear too dark.

Too low a kilovolt potential, and the image will lack detail and be too light. Some small seeds need to be x-rayed at a low kilovolt potential to give the correct penetration but need a long exposure to produce a radiograph with enough density to provide good contrast. Xray inspection cabinets are manufactured that operate in this very low kilovolt potential range for examining small items such as seeds. They are designed for total protection of the operators, with complete lead shielding and safety interlocks on the door.

X-radiography has proven useful for studying the seeds from many wild species, which often can be empty or poorly formed. Some laboratories test every lot of seeds they receive with x-rays and get a good initial evaluation. X-radiography can be of great value in evaluating germination test results, because it is much faster than cutting open seeds that failed to germinate.

Empty seeds will never germinate, and damaged or poorly developed seeds will seldom germinate. The excised embryo or tetrazolium test for difficult-to-cut seeds can be speeded up by x-radiography. The seeds are first placed on the x-ray film or paper in a manner that will allow the comparison of the exact image to the exact seed. This is done by placing the seeds on an additional piece of paper before placing the paper on the x-ray film or paper.

If orientation of the seed is important, as in double-seeded fruits such as dogwood, the seeds can be placed on adhesive tape and that then laid on the paper. The seeds should be oriented so that both seeds in the fruit can be viewed and the tape prevents them from turning. After the radiograph is made, the seeds are gently slipped off the x-ray paper so that the seeds are kept in order for cutting. Only those seeds that are morphologically sound in the radiograph need to be cut.

Other Quick Tests

As stated in the introduction, cutting tests are very limited in their application. However, they can provide useful information on full seed percentages and the condition of the internal structures. For example, color of the tissue cannot be determined in a radiograph, which is only black and white. Seeds that are cut and found to be dark are not likely to germinate. New and unfamiliar images in a radiograph require cutting the seed to determine what is actually in the seed. With slash and longleaf pines, cutting can reveal embryos that have initiated chlorophyll (turned green), a result of a seedlot having been held too long at high moisture. This is a sure sign of a weakened seedlot.

Hydrogen peroxide (H_2O_2) has been used as a quick test for western conifers. In this test, the seeds are floated in solution of 1.0% hydrogen peroxide overnight. The radicle ends are then clipped and the seeds incubated in the dark at 20 to 30 °C for 10 to 12 days. Counts of germinates are made at 3 to 4 days and at 10 to 12 days. The hydrogen peroxide solution is changed at the first count.

SOWING RATES

A sowing rate is the amount of seeds sown in a unit area of nursery bed to produce the desired number of seedlings. The following formulas show how seed test data are used to determine this rate.

Weight of seeds to sow in a nursery bed (width x length) is equal to

$$\frac{\text{bed width x bed length x seedlings desired per area}}{\text{germination x seeds per weight x purity x survival factor}}$$

Number of seeds needed to sow per area of nursery bed is equal to

germination x purity x survival factor

In both of these formulas, the survival factor is the ratio of the number of seedlings expected to the number of viable seeds planted. It is derived from experience in the given nursery and should be constantly updated with new information collected from history plots. History plots are permanent sample plots in a nursery bed used for carefully monitoring the number of seeds sown and the number and quality of seedlings produced. For example, if 100 seeds are sown on a square foot, germination is 80% in the laboratory, and 60 seedlings actually grow on the square foot, then the survival factor is 60/80 or .75 (75%).

Computing sowing rates for containers is somewhat different, because we must predict the probability of an empty cell in the container. The probability that a container cell is empty is equal to 1 minus the probability that at least 1 seedling is in the cell. Sowing 1 seed/cell, this probability is 1 minus germination. With a 90% germination, the probability of an empty cell following single- seed sowing is 0.1. In sowing 2 seeds/cell, the probability of no seedling in a cell drops to .01, but now there are 81 cells (.90 x .90) out of 100 that will have 2 seedlings/cell.

Double seedlings per cell requires thinning to 1 seedling/cell for proper growth. Thus, in container nurseries it is necessary to choose between empty spaces and thinning. In this example, to go from 10 empty cells to 1 empty cell/100 cells, 81 seeds were wasted. For 10,000 seedlings, 8,100 seeds would be wasted which would be 112 to 224 g (3.9 to 7.8 oz.) of seeds when the seeds per kilo is 72,300 (32,900 seeds/lb.). Thinning also requires more labor and may be dysgenic by favoring early germinating genotypes. Sowing extra containers is another strategy followed to compensate for empty cells. The empty cells are still present but enough seedlings are produced without the problems of thinning.

TEST LIMITATIONS AND VARIATION

Standardised laboratory tests are designed to give maximum values with minimum variation, allowing the results to be repeated. Without repeatability, there would be no standard by which to compare different seedlots. Assessments would become even more difficult if seed tests were conducted in random fashion to mimic field conditions. Although the correlation between laboratory and field germinations is frequently low, experience has shown that seedlots with higher germination scores will, over time, give more germination in the field. Further improvement in predicting field performance may result from improvements in the techniques of vigour testing.

Because seeds are biological, they usually are quite variable in size and performance. Natural things simply show more variation than is usually seen in manufactured items. Test results, therefore, can vary more than perhaps is expected and still be accurate.

SCHEDULING SEED TESTS

Tests are needed to formally determine the quality of the seedlot upon completion of the conditioning process. All variables then need to be measured: germination, seed weight, purity, and moisture content. Moisture needs to be monitored during storage to be sure it is being properly maintained. If there are no changes in moisture content, then seed weight and purity will not change and viability will change very slowly, if at all.

Some annual monitoring of seed moisture is necessary to ensure that storage conditions are being adequately maintained. The viability can be retested at 3- to 5-year intervals with a current test always done no more than 6 to 9 months before sowing. If a longer time passes before sowing, some deterioration could occur, resulting in changes in germination. Determining viability in some seeds takes a long time, and thus it is important to schedule adequate lead time into the production schedule.

COMMERCIAL TRADE OF TREE, SHRUB, AND NATIVE PLANT SEEDS

Official rules for testing seeds developed by the International Seed Testing Association or the Association of Official Seed Analysts are very important to the orderly buying and selling of seeds. These rules give standardised procedures that can be repeated with acceptable variation no matter what laboratory conducts the test. In addition, the seed testing associations conduct comparative tests among themselves to verify that the procedures are being applied uniformly and within tolerable limits of error.

Such a system is important: it assures sellers that they are offering good seed for sale, and it reassures buyers by giving them reliable information on which to base their purchases. There are also consumer protection seed laws in some countries or states requiring that seeds offered for sale meet certain minimum standards or be accurately labeled as to their quality. Without a repeatable system of testing procedures, such laws would be impossible to enforce. Both the consumer and the reputable seed dealer would suffer. Uniformity of the rules combined with uniformity in applying the rules equals order in the marketplace.

REFERENCES

Bonner F.T., "Determining seed moisture in *Quercus*", *Seed Science and Technology*, 1974.

Ching T.M, Parker M.C., "Hydrogen peroxide for rapid viability tests of some coniferous tree seed", *Forest Science*, 1958.

ISTA [International Seed Testing Association], "International Rules for Seed Testing", *Seed Science and Technology*, 1996.

ISTA, *Handbook of vigour test methods*, 1995.

Piotto B., "Effects of temperature on germination of stratified seeds of three ash species", *Seed Science and Technology*, 1994.

4

Production of Onion Seeds

Onions originated in Iran and Pakistan and were established staple foods of Egypt and India around 1500 B.C. By the middle ages, they were in Europe and brought to North America by Spanish settlers.

The unique flavour and odor of onions have made them an excellent food source that provides zest to many dishes. Moreover, the numerous ways in which onions are prepared from boiling, frying, stewing, baking, pickling, to eating raw make them a versatile food source. The recent popularity of a health conscious world for salad bars has further increased their agricultural importance.

Nutritionally, onions are low in calories (approximately 40 calories for an average size onion) and high in ascorbic acid. They are classified into three agriculturally important groups: common, aggregation, and proliferous. The common onion is the most agriculturally important group. It is a bulb producing type reproduced by seed. The aggregation group is characterised by formation of many lateral bulbs or shoots and is reproduced vegetatively. An example of this is shallots. The proliferous group has bulblets form on the inflorescence and is vegetatively reproduced. An example of this is top-set onions.

PLANT DEVELOPMENT

Vegetative. Onions are a member of the Liliaceae family and are considered a cool-season crop. They are herbaceous biennial plants. The first year of development results in the establishment of a fibrous, shallow root system attached to a base plate of fleshy leaves arising around a small conical terminal stem, a structure botanically termed a bulb. Because the root system is shallow, onions are particularly prone to dehydration injury. The leaf sheaths that project above the ground are sometimes called a false stem to distinguish them from the true stem at the bulb's base. Bulb formation is initiated by changes in day length. Very early types require a 12 hour photoperiod and later types initiate bulbing with a 15 hour photoperiod.

Reproductive. Flowering and seed formation occur in the late Spring or early Summer in response to a vernalisation treatment during the Winter. The onion inflorescence is an umbel that produces 50 to 2,000 flowers. Flowering can be as long as two weeks and is not uniform since the umbel actually consists of aggregations of smaller 5 to 10 flowered inflorescences called cymes. From one to over a dozen flower stalks can be produced on a single plant dependent on its development. More flower stalks are produced on plants grown from bulbs than those grown from seed.

The stigma of the flower becomes receptive to pollen after the anthers have released their pollen thereby reducing self-fertilisation. However, open-pollination primarily by insects such as honey bees and leafcutter bees carrying pollen from one flower to the next is common. Optimum seed production occurs when 12 to 24 hives per hectare (5 to 10 hives per acre) of hives are moved to the edge of seed fields. In some years, when dry and hot weather prevails, flower nectar production becomes concentrated and unattractive to bees. In these instances, the bees move to other crops and pollination does not occur.

High temperatures also lead to death of flowers and seed abortion. To minimise this problem, misting of onion seed crops by overhead sprinklers increases seed yields when temperatures exceed 38°C (100F). In 1925, male sterility in onions was discovered and this finding quickly led to the development of a hybrid onion seed industry.

In the production of hybrid seed, the ratio of female to male rows varies but usually is 8 to 2 or 10 to 2. For pollination to be successful, the female and male lines must flower at the same time. Two approaches are used to ensure that this occurs: 1) the parent lines can be planted at different times or bulbs can be harvested and 2) differences in storage temperature can alter the time of flowering after replanting of the bulbs in the Spring. After pollination, the male plants must be removed by mowing or disking to ensure that seed heads from the male and hybrid do not become mixed during harvest as a consequence of lodging.

SEED PRODUCTION

The seed is three sided, 3.0 to 3.5 mm (0.12 to 0.14 inches) long. Its surface is wrinkled, dull and greasy looking, black, and has a depressed scar at one end. Within the seed, the embryo is 6.0 mm (0.24 inches) long and 0.4 mm (0.02 inches) in diameter and is curved in a complete circle. Most of the embryo is comprised of a single cotyledon to which is attached a very small shoot apex, first foliage leaf; and short primary root. It is embedded in a tough thick-walled endosperm. The 1,000 seed weight of onion is 3.6 g (126,000 seeds per pound).

Dry onions are produced throughout the United States. However, the climatic conditions for onion seed production are more demanding than those for bulb production and are concentrated in the dry areas of the West. Most of the onion seed produced in the United Sates comes from the Snake River Valley of Idaho and Oregon, the Willamette Valley and Madras regions of Oregon, the Columbia Basin of Washington and the Salinas Valley of California. In 1991, 1.6, 3.2, and 10.5 hectares (4, 8, and 26 acres) of foundation, registered, and certified seeds were produced, respectively.

Onion seed production can be open-pollinated or hybrid. Since hybrids have greater uniformity, increased yields, and greater disease resistance, most seed produced today is hybrid. These can be produced either by a bulb-to-seed or seed-to-seed method. The bulb-to-seed method is when onion bulbs are grown, harvested in the fall, stored, and replanted in the Spring. The cold vernalisation requirement is experienced during bulb storage in the Winter. This method is more expensive than the seed-to-seed method but it also permits a grower to easily discard off-types, diseased or otherwise undesirable bulbs. It is practiced primarily for maintenance of onion stock seed.

The seed-to-seed method is when seed is planted in the Summer, carried over through the Winter as bulbs in the field, followed by flowering in the Spring. This method results in higher seed yields than the bulb-to-seed method because there are more plants per hectare (acre) and less time is invested in bulb establishment. While overall costs are less than the bulb-to-seed method, careful attention must be given to pest control. Roguing programs to eliminate off-type plants are also required. Because this method is the most common approach to onion seed production.

Tillage

An onion field should not be used for seed production more than once every four years. The field must be free of perennial weeds and soilborne diseases. The soil should be friable, fertile, well supplied with humus, and well drained. The soil should be plowed to a depth of 15 to 20 cm (6 to 8 inches), harrowed, and worked until a uniform firm seedbed 10 to 15 cm (4 to 6 inches) deep is established.

Planting

Onions seeds are planted 1 to 3 cm (1/2 to 1 inch) deep in rows 50 to 75 cm (20 to 30 inches) apart dy drills at a rate of 4.5 to 6.7 kg per hectare (4 to 6 pounds per acre). Higher seeding rates and closer row spacings have, in some cases, led to higher yields. Planting should occur as soon as the danger of frost is over. If the fields are irrigated or rainfall is common, row spacings of 30 to 45 cm (12 to 18 inches) can be used.

Fertilisation

The objective of fertilising onion seed crops is to ensure that the onion bulbs reach adequate size before the onset of Winter in order that they can produce vigorous seed stalks the following Spring. Onions respond well to fertilizers because the root structure is limited and shallow. They grow best on soils that are slightly acidic (pH 6.0 to 6.5). Mineral soils should be amended with organic matter to assist in moisture retention and improve the soil physical structure.

The equivalent of 56 kg per hectare (50 pounds per acre) of nitrogen should be residually available in the soil or the equivalent applied broadcast before planting. Additional applications of nitrogen per hectare (acre) can be applied in one or two sidedressings to make it easily available to the roots. Both phosphorus and potassium should be broadcast at the rate of 67 kg per hectare (60 pounds per acre) prior to planting. Later applications, if needed, can be sidedressed. Fertilisation of nitrogen and potassium is not recommended at flowering because these make the flower nectar unattractive to bees.

Minor element deficiencies occur in onion, particularly on muck soils. Copper deficiencies result in thin, poorly colored scales that have a poor bulb storage life. This can be corrected with the application of 224 kg per hectare (200 pounds per acre) of copper sulfate. This treatment lasts for several years. Magnesium deficiencies result in leaf chlorosis and are often encountered on muck soils that are alkaline or have recently been limed. Manganese deficiencies cause a gradual interveinal yellowing, twisting, curling, and stunting of leaves. Manganese sulfate at 168 kg per hectare (150 pounds per acre) mixed with the broadcast fertiliser alleviates this problem.

Irrigation

Seed onions produced in the dry West require irrigation because of their shallow roots and high demand for water. An onion crop can use 1 1/2 to 2 1/2 acre feet of water during its development and maturation. Soil moisture should always be above 65% field capacity. In areas where irrigation is practiced, seedbeds are usually irrigated immediately after planting. Then, one to three irrigations are necessary after seedling emergence to ensure rapid, continued growth. As the plants initiate bulbing, irrigation is stopped and the soil allowed to dry.

Weed and Pest Control

Onions are poor competitors against weeds because of their slow growth, small stature, shallow roots, and lack of dense foliage. Weeds interfere with harvest and can become contaminants of the crop seed. The most effective weed control measures include crop rotations, cultivation, and use of selective herbicides. A short fallow period before planting also permits an opportunity to use tillage and non-selective herbicides to control weeds.

Cultivation is one of the most successful weed control techniques in onion. Weeds can be destroyed during seedbed preparation. After planting, cultivation every one to two weeks until bulbing occurs is useful. This practice not only suppresses weeds but also loosens the soil after it has hardened following irrigations and rains. Initial cultivations should be shallow since the onion roots are close to the soil surface and damage to the crop can occur.

Winter annuals, summer annuals, and perennials are troublesome weeds in onion fields. Winter annual weeds are London rocket, shepherdspurse, sowthistle, prickly lettuce, and wild oats. Summer annuals include barnyardgrass, yellow mustard, lambsquarters, pigweed, purslane, and sunflower. Persistent perennial weeds are yellow nutsedge, quackgrass, Canada thistle and field bindweed.

Onions are subject to many of the same fungal and insect pests that invade other crops. Scape and umbel blight caused by Botrytis allii infects onion seed stalks just below the umbel causing it to be girdled and topple which reduces seed yield and quality. Downy mildew (Peronospora destructor) is a problem when cool, wet weather prevails during flowering. It causes chlorotic lesions on the leaves and seed stalks. Pink root caused by Pyrenochaeta terrestris is a soilborne pathogen that infests onions at any stage of development and causes the roots to turn pink followed by brown to black and then death.

It is suspected when the above ground portion of the onion plant shows stunting and reduction in vigor. Fusarium basal rot (Fusarium oxysporum) is detected when plants show a progressive yellowing and die back from the tops of the leaves. Infected plants have roots that are dark brown, flattened, hollow, and transparent and the stem plate is brown, rotted, and discolored. This disease reduces seed yields.

There are two common insect pests of onions: thrips and maggots. Onion thrips (Thrips tabaci) feed on leaf surfaces during warm, dry weather causing the leaves to turn white or silver. They also feed on the flowers of seed crops causing reduced seed set and smaller seeds. When thrip damage is suspected, it is best to examine the angles of leaves since this is a common hiding place for these small insects. Control of thrips is difficult in seed crops because use of insecticides can simultaneously harm pollinating bees. Onion maggots (Delia antigua) are the larvae of a small fly that lays its eggs in the soil. The eggs hatch to produce the larvae which attack the onion plants at any age but do most damage in the Spring by invading germinating seeds and seedlings. Seedlings are initially stunted, then yellow and die.

Harvesting, Drying, and Threshing

Although all seed heads on an onion plant do not mature simultaneously, there still is usually one harvest in the seed field. This is accomplished by harvesting the seed heads at about 30% moisture (the heads have some opened capsules with black, ripened seeds exposed) by hand cutting 10 to 15 cm (4 to 6 inches) of the flowering stem below the umbel. When cutting, the umbel is supported in the palm of the hand and held between the fingers to avoid seed shattering. The seed heads are placed into a burlap sack, loaded onto trucks, and hauled from the fields.

They should not be left in the sacks for more than 1 to 2 days since heating of the mass occurs which is detrimental to seed quality. The heads are dried by forced air in boxes or large bins or by spreading in a shallow layer on a clean surface exposed to the sun. The seeds are dry enough for threshing when the capsules and small seed stems are brittle and readily break when rolled in the palm of the hand. If seeds are dried with heat, 32°C (90F) until the moisture content is below 18%, then at 38°C (100F) until the moisture content is below 10%, and at 43°C (110F) until the storage moisture content is attained. Threshing is done with a combine. Yields range from 560 to 784 kg per hectare (500 to 700 pounds per acre) for open pollinated lines and 336 to 1,120 kg per hectare (300 to 1,000 pounds per acre) for hybrid lines.

Conditioning

After threshing, onion seeds are conditioned to eliminate weed seeds, light seeds, and chaff. Initial cleaning is with an air-screen cleaner followed by a gravity table. In some cases, the light separated seed is washed. Heavy, good seeds sink while light, poor seeds float. The washing treatment should not exceed three minutes since the seeds rapidly absorb the water and quality is subsequently reduced. The heavy seeds are retained and immediately spin dried in a centrifuge and further air-dried. The seed is dried to a moisture content of 12% or less prior to storage.

Storage

Onions possess one of the most rapidly deteriorating seeds among major crops. They quickly lose complete viability in less than a year when stored under hot, humid conditions. If properly stored, however, viability of onion seed can be retained for long periods and low seed moisture content and low storage temperature favor longer storage life. Of these two parameters, seed moisture content is the easiest to adjust. Onion seeds stored at 6.0% moisture content have retained viability up to three years. Small lots of valuable breeding onion seed material are often stored in desiccators over calcium chloride. Under these conditions, their moisture content will equilibrate to about 2.0% and germination can remain high for as long as nine years in storage.

REFERENCES

Comin, D., *Onion Production*, Orange Judd Publishing, New York, 1946.

Jones, H. A., and L. K. Mann, "Onions and Their Allies", *Botany, Cultivation, and Utilization*, Leonard Hill, New York, 1963.

Rabinowitch, H. D., and J. L. Brewster (eds.), *Onions and Allied Crops*, CRC Press, Boca Raton, 1990.

5

Production of Cabbage Seeds

Most wild forms of cabbage are found along the coasts of the Mediterranean Sea which is considered their point of origin. Greek and Roman civilisations were the first to recognise the value of cabbage. Its ease of growth and excellent storability ensured its place as an important crop. Initially, these cultures used cabbage for medicinal purposes. Cabbage juice was employed as a gargle against hoarseness and the leaves covered wounds and ulcers to hasten healing. Later, cabbages were eaten raw as well as cooked. By the time the Greek and Roman empires expanded into Europe, cabbage was considered a staple that was carried with them and introduced throughout the continent. European settlers brought cabbages in the 16th and 17th centuries to North America. By the 1700s, cabbage was a common crop grown by the colonists.

There are three major types of cabbage classified according to market demand: fresh market, storage, and sauerkraut. Fresh market types usually form conical to globe heads and have a range of maturities from 50 to 95 days. Storage types are slow in growth and have a firm head with thin, finely veined leaves. Sauerkraut types form large heads that mature in about 130 days. Cabbages are used in a variety of ways from raw in salads and soups, boiled or steamed in mixtures of vegetables and cabbage rolls, to pickled in sauerkraut. Nutritionally, cabbage is a good source of vitamin A and possesses moderate amounts of vitamin C and the B-complex vitamins. It also has high levels of potassium and some calcium and phosphorus.

PLANT DEVELOPMENT

Because cabbage is a biennial plant, the first year of growth requires the production of a vegetative head followed by flowering and seed production in the second year. It is considered a cool-season crop which grows best between temperatures of 10 and 25C (50 to 77°F).

Vegetative. The cabbage seedling often produces a red hypocotyl with two notched cotyledons and a tap root with lateral roots. The first three leaves have a petiole while later leaves are completely sessile. The leaves are unique from many other crops because they are coated with a layer of wax. The first leaves unfold normally but later leaves only partially unfold and do not expand as much as earlier leaves ultimately forming a head that varies from flat-topped to long-oval in shape.

Reproductive. Cabbage is not photoperiodically sensitive but the apical meristem of plants that have begun to form a loose head must be exposed to four to six weeks of 4 to 7C (39 to 45°F) for vernalisation to occur eliciting production of a reproductive apical meristem. Vernalisation of the seed or seedling does not cause flowering. To permit the seed stalk to penetrate the tightly oppressed leaves of the cabbage head, the head is often quartered by a knife. The seed stalk then attains a length of 1 to 2 meters (39 to 78 inches). The inflorescence is a raceme on which are born four petaled, yellow flowers typical of the Brassicaceae.

The flowering process starts from the bottom of the raceme and moves to the top. This often results in racemes on which mature siliques are found on the bottom with unopened flower buds on the top causing an uneven setting and ripening of seed on the plant. The flowers are insect pollinated. When hybrid seed is produced, this normally occurs in a 2:2 ratio of male:female parent rows. After pollination, elongation of a dry pod called a silique containing 10 to 30 seeds is completed after three to four weeks. Dehiscence occurs when the two silique valves split from the point of attachment to the plant upward leaving the exposed seeds attached to the placenta.

SEED PRODUCTION

The cabbage seed is almost round, 2.0 to 3.0 mm (0.08 to 0.12 inches) in diameter, reddish brown to grey to black, with a discernable radicle groove and hilum. The seed coat possesses a small reticulated surface. The embryo is large, folded and little endosperm is present in the mature seed. The 1,000 seed weight is 3.6 g (126,000 seeds per pound).

Cabbage is produced throughout the United States with Texas, Florida, California, and Georgia leading the production of fresh cabbage and New York, Wisconsin, and Ohio producing the largest quantity of cabbage for sauerkraut. Seed production in the United States occurs principally in the Skagit Valley of the Puget Sound in Washington state. This area is not only noted for its mild climate and soils favorable for cabbage seed production, but also the abundance of syrphid flies and bees which are excellent pollinating insects.

Tillage

Cabbage requires an abundant supply of water. As a result, it typically grows best on soils that are heavy and contain a high level of organic matter. The pH should be between 6.0 to 6.5 for optimum growth. The planting bed for cabbage should be deep and well worked to enable rapid seedling root development and access to water. This can be accomplished by disking the soil. Disk cultivators are useful in preparing seed bed rows about 7 to 10 days before planting because they further pulverise the soil. Additional smoothing of heavy soils, if required, can be done with a harrow.

Planting

Normal cabbage seed production begins with planting in the Summer followed by the half-grown plants over wintering, flowering in the Spring, and maturation and harvest of the seeds in the Summer. All stock seed is tested or hot water treated for black rot prior to planting. Cabbage is planted initially in beds or the greenhouse. The seedlings are then transplanted 45 days later in the field in rows 107 cm (42 inches apart) with a spacing between plants of 40 cm (16 inches) to produce a population of 3,645 to 4,050 plants per hectare (9,000 to 10,000 plants per acre).

Hybrid seed production is more labor intensive. Each parent is normally staked and stringed to allow improved pesticide spraying and facilitate subsequent hand cutting of the inflorescence. The stakes and strings insure that the parents stay separated. The nick between female and male parents must be managed to minimise the number of inbreds and maximise the number of hybrids produced. To accomplish this, bees are placed in the field at 5% bloom. In some cases, plants are topped if one parent comes into bloom earlier than the other.

Fertilisation

Nitrogen during the establishment of cabbage plants leads to a softening of tissues and an increased susceptibility to frost. As a result, nitrogen is usually omitted at planting with phosphorus and potassium applied at rates of 100 and 150 kg per hectare (90 and 134 pounds per acre), respectively. If nitrogen is necessary, it is split as a sidedress in the Fall and added in the Spring just before bolting. This ensures maximum growth of the sidebranches which enhances seed yield. Too much nitrogen in the Spring or Fall enhances soft tissue development, increases the risk of lodging, and promotes extended flowering which may cause increased inbreds in hybrid seed production. Cabbage is not as sensitive to micronutrient deficiencies as are other members of the Brassicaceae.

Weed and Pest Control

Chemical and mechanical weed control programs are practiced with cabbage. Seedbed preparation should eliminate most weeds. When weeds are persistent, however, a preplant or preplant incorporated application of herbicides can be used. The extensive and shallow rooting system of cabbage limits deep cultivation to control weeds, particularly when the soil is hard. Only when weeds are present should large plants be cultivated and all cultivation should be discontinued after heading to avoid further plant damage. The most serious weeds include tiny vetch, wild geranium, and wild turnip which must be controlled or removed prior to harvest.

The cabbage plant is subject to a host of diseases. The most serious of these are downy mildew and alternaria which must be controlled up to harvest. This is accomplished by spraying the hybrid seed fields 3 to 5 times with appropriate fungicides. Other diseases include black rot, yellows, club root, and blackleg. Black rot (Xanthomonas campestris) first appears as a yellowing of leaves and blackening of veins. Later, the plant becomes dwarfed and the heads one-sided.

Yellows (Fusarium oxysporum f. conglutinans) occurs in plants grown in warmer climates and is indicated when the plants turn yellow, the leaves fall off, and the plants die. Club root (*Plasmiodiophora brassicae*) produces malformed, club-like roots. Blackleg (Phoma lingam) attaches to the stem of young plants causing dark shrunken areas that leads to plant wilting and death. Black rot and blackleg are diseases carried on or in the seed. These can often be hot-water treated to eliminate the pathogen(s). Care must be taken not to kill the seed in the hot water treatment. Seed treatments, where approved, can also be very effective against these diseases.

Cabbage is susceptible to an array of insects and other pests. Cabbage aphids and pod borers are the most serious insect pests of cabbage seed production. These sucking insects possess a protective waxy coating similar to cabbage leaves. Insecticides must contain an appropriate detergent or sticker to control these pests. Other insects include cabbage worms and the cabbage looper. The harlequin bug is another sucking insect that is about 1.0 cm (0.4 inches) long and is mottled red, black, or yellow in appearance. Root knot in cabbage is caused by a parasitic eelworm that produces irregularly shaped galls on the root that sometimes can be confused with club root. Rotations or soil fumigations are the best control measures for this problem.

Irrigation

Cabbage demands a high and consistent quantity of water for optimum development. This is attributed to its large leaf surface area and resultant loss of water through evapotranspiration. The soil water level should be maintained at 60% field capacity or higher throughout crop growth. After bolting, irrigation should be discontinued.

Harvesting, Threshing, and Drying

Generally, each parent is harvested separate from each other. However, in some cases when both parents are stongly self incompatible, they are harvested together. The uneven maturation of siliques on the cabbage raceme and the strong tendency for the seed pods to shatter creates harvesting problems for obtaining maximum yields. For these reasons, harvesting of hybrids is done by hand by cutting the seed stalks with a knife or shears.

Open pollinated varieties are swathed by machine. In both cases, this is started when the basal siliques turn yellow and the enclosed seeds are brown. Harvesting in the morning after a dew tends to reduce the level of seed shattering. The seed stalks are placed in windrows for continued drying or on canvas to collect the fallen seeds from dehisced siliques. After the plants are dried for 10 to 14 days, they are threshed in a combine. Combine cylinder speed and concaves should be adjusted to minimise damaged seed. Cylinder speeds should not exceed 700 rpm because of the tendency for dried seeds to easily split.

Conditioning

Removal of remaining chaff and straw from the cabbage seed can be done on an air-screen cleaner. Separating splits from whole seeds is accomplished with a spiral separator. In addition, cabbage seed is normally sized into several sizes for better precision planting.

Drying and Storage

Cabbage seed can remain germinable for four to six years if properly dried and stored at a low moisture content in a low relative humidity (50% or less) environment. The seed should be at a moisture content not exceeding 6% at storage. When drying is necessary, the temperature should not be above 45C (113°F) with lower moisture seeds more resilient to higher temperatures.

REFERENCES

Nieuwhof, M., "Cole Crops", *Botany, Cultivation, and Utilization*, Leonard Hill, London, 1969.

6

Production of Lettuce Seeds

The exact origin of lettuce is unknown. Paintings of a cos-like lettuce are found in Egyptian tombs constructed about 4500 B.C. From Egypt, it is believed that lettuce quickly spread around the Mediterranean basin and was adopted as a popular vegetable by the Greek and Roman civilisations. The Romans further developed differing lettuce types that were broad-leaved, nonheading, nonspiny, had decreased latex content, and were resistant to early seed stalk formation. Eventually, lettuce spread throughout Europe where the present six types of lettuce were selected and produced. It was brought to the New World by the Spanish settlers and was marketed as a vegetable crop in seed catalogues by 1806. By the 1880s, over 20 lettuce varieties were available for production in North America.

Lettuce is unique among vegetable crops because it is used exclusively as a fresh, unprepared component of salads. There are six morphological types of lettuce: crisphead, butterhead, cos, leaf, stem, and Latin. Crisphead lettuce, often called iceberg, is most common in the United States. This type of lettuce produces large, heavy (up to 1.0 kg/2.2 pound) heads with tightly folded, green outer and white or yellowish inner leaves. Butterhead lettuce forms relatively small, loose heads with broad, oily, crumpled, soft-textured leaves and is commonly consumed in northern Europe. Cos lettuce, often called romaine, has elongated leaves that are dark or light green with heavy ribs forming a loaf-shaped head and is most popular in southern Europe and the Mediterranean basin.

Leaf lettuce produces a variety of leaf types and color but all are characterised by forming a rosette of leaves that do not form a head. This lettuce is used primarily in home gardens because of its ease of production. Stem lettuce is one of the few lettuce types that is not consumed raw. It produces leaves that are coarse, peeled, and cooked. This lettuce is grown primarily in the Orient. Latin lettuce resembles butterhead lettuce in forming a loose head but the leaves are elongated and similar to cos lettuce. This lettuce is popular in the Mediterranean countries and South America.

Lettuce is relatively low in essential nutrients compared to other vegetables. It is known for its low calorie level which is attributed to a very high (94 to 95%) water content per leaf. Lettuce contains moderate amounts of vitamins A and C and calcium as well as some phosphorus, iron, sodium, and potassium. Despite its low nutritional value and lack of diversity in food preparation, lettuce is so important as a main component of salads that it ranks fourth in vegetable popularity behind only tomato, citrus, and potato.

PLANT DEVELOPMENT

Like sunflower, lettuce is a member of the Asteraceae family. The cultivated lettuce is closely related to and can easily cross with the common wild lettuce (*Lactuca serriola*). Both produce high levels of latex in the stem and leaves and this trait is responsible for the derivation of the generic name Lactuca meaning "milk-forming". Lettuce is a cool-season crop that requires abundant moisture and ample sunlight for optimum development. Cool nights are important to establish normal head development and high temperatures promote seed stalk formation.

Vegetative. Lettuce is an herbaceous annual that differs from its wild-type relative by producing a rosetted, short-stemmed vegetative plant. The leaves vary widely in shape from round to obovate with varying degrees of cupping and indentation. The root system develops a deep, penetrating taproot with the major absorbing surface being in the top 30 cm (12 inches) of soil.

Reproductive. Lettuce varieties are either long-day sensitive or day-neutral and high temperatures (above 18C/64°F) promote bolting (a change from the vegetative state to flower formation). In most cases, the seed stalk or inflorescence rapidly emerges from the head or leaf structure. In some cases, particularly with crisphead lettuce, the head is so tight and firm that the seed stalk is unable to emerge from the head. To enhance seed stalk emergence, three methods are used:

1) *deheading*: physical removal of head leaves surrounding the seed stalk,
2) *slashing and quartering:* cutting the seed head at its apex in an "x-shape" that loosens the tight-fitting leaves over the seed stalk,
3) *application of growth regulating chemicals:* application of gibberellic acid (20 to 500 ppm) to lettuce leaves at the three to five-leaf stage promotes bolting.

Lettuce produces a panicle with many individual compound flowers formed into a capitulum inflorescence. As a result, many flowers are formed at differing stages of development. Over 90% of the seed yield comes from flowers that open during the first 35 days of the 70 day flowering period. Seeds produced during this time are bigger and heavier than those produced later. The flowers are yellowish or white yellow and consist of several florets borne on a flat, naked receptacle. The corolla rays are five toothed and truncate at the end. The flowers contain fused stamens that release pollen at the time of style elongation through the anther sheath ensuring self pollination. Flowers open only once for a short period in one morning and then never reopen; a process that also reduces access to foreign pollen.

SEED PRODUCTION

The lettuce fruit (seed) is an achene. It is lanceolate, flat, and the cream- to back-colored fruit wall has about 8 lengthwise ridges. The achene is 4 to 5 mm (0.16 to 0.20 inches) long, beaked, and topped with a number of large soft, capillary, white or brown, delicate pappi (Figure 00.7). The 1,000 seed weight of lettuce seed is 0.6 to 1.0 g (453,000 to 755,000 seeds per pound) dependent on cultivar.

Approximately 70% of the lettuce seed in the United States is produced in California. This is due to the availability of favorable soils, climate, dependable water supply, and access to available labor since this crop requires a high labor input.

Tillage

Lettuce can grow on soils that range from fine sandy loams to clays to mucks but they must have a high organic matter content and adequate nutrients available in the upper 25 to 30 cm (10 to 12 inches) of soil. Since calcium is an important element for rapid leaf growth, the soil should be at a pH of 6.0 to 6.5 to ensure its availability.

Typical vegetable tillage practices should be followed. The soil should be plowed to 30 cm (12 inches) as early in the season as possible. The seedbed is then smoothed by disking or harrowing and then leveled with a spiked harrow. When irrigation is necessary, elevated seed beds are prepared and then dragged flat.

Planting

High quality lettuce seed should be used in planting for greater uniformity of growth. It is often pelleted to facilitate precision planting. The minimum soil temperature required for germination of lettuce seed is 1.7C (35°F) and the optimum temperature is 24C (75°F). High temperatures of 33 to 35C (91 to 95°F) inhibit germination. When lettuce is planted under high temperatures, it is often planted at night and irrigated to induce germination before the high day temperatures are experienced.

The seeds also can be osmoconditioned in growth regulators such as kinetin and gibberellic acid and redried, or pregerminated in a gel then planted by fluid drilling. Lettuce seed is planted 1 cm (0.4 inches) deep at a rate of 2.4 kg per hectare (2.0 pounds per acre) and spaced 5.1 to 7.6 cm (2 to 3 inches) apart in rows 38 cm (15 inches) apart. The plants are thinned four to six weeks after emergence to 25 to 30 cm (10 to 12 inches) apart to give a population of 74,000 plants per hectare (30,000 plants per acre).

Fertilisation

On mineral soils, nitrogen is usually limiting and should be applied at planting, after thinning, and at bolting. Phosphorus and potassium should be applied at planting because of the crop's inability to extract these elements from deeper portions of the soil. Muck soils have a lower nitrogen requirement but require phosphorus and potassium. Thus, while it is difficult to generalise on fertiliser application rates, a ratio of 3-2-2 usually provides suitable crop growth. Other elements do not often limit lettuce growth. However, under high or low moisture situations, a condition known as calcium-related leaf tip burn (similar to blossom-end rot in tomatoes) occurs.

Weed and Pest Control

Crop rotations with tomatoes, cucurbits, sweet corn,

spinach, beets, and carrots assist in minimising the buildup of lettuce weeds and diseases. Weed control is accomplished by both cultivation and herbicide application. Since herbicides are increasingly regulated, particularly in California, good cultivation practices should be encouraged and practiced. These begin with proper land preparation before planting.

After planting, if weeds are present, a shallow cultivation not more than 5.1 to 7.6 cm (2 to 3 inches) deep may be necessary. This practice cuts the weeds at or immediately below the soil surface, leaves a shallow mulch on the soil, and protects the small developing roots of the lettuce plant from mechanical damage. It may still be necessary for hand weeding between rows at later stages of crop development. Common weeds of lettuce fields are lambsquarter, pigweed, henbit, chicory, groundsel, sowthistle, and chickweed.

Lettuce is prone to a host of pathological disorders. From the perspective of the seed producer, lettuce mosaic virus is the most serious of these because the virus is carried in the seed. Infected seeds germinate to produce plants that are stunted, have yellow leaves, and often fail to produce a harvestable head. More importantly, sap sucking insects such as aphids feed on infected plants and serve as vectors for dissemination of the disease to healthy plants throughout the field. Seed production on infected plants can be reduced by 68% and seed weight by 62%.

The virus is transmitted to 1 to 15% of the seeds maturing on the infected plant. As a result, basic seed stocks are often produced in insect-free structures or areas where temperatures are too high for aphid attack. Most major lettuce producing areas require planting only of mosaic-indexed seed set at some local arbitrary level such as 0.5%. This level is determined by seed companies planting the seeds in flats in greenhouses and growing the seedlings to the three to four leaf stage. Infected seedlings are stunted and lighter than healthy seedlings. The use of mosaic-indexed seed has been very successful. Estimates of reductions in losses of 95 to 100% have been reported as a result of this practice

Other common diseases of lettuce are found. Downy mildew occurs on lettuce plants grown at low temperatures and high humidity. Symptoms appear first on the lower leaves as large yellow to brown lesions. Sclerotinia rot is common under cool, moist growing conditions or when the crop has been excessively irrigated on heavy soils. It is characterised by an initial wilting and collapsing of lower leaves followed by the entire head which eventually rots at the base. Botrytis or grey mold can cause damping-off of seedlings or head rotting in more mature plants. Powdery mildew occurs under warm, humid conditions and is evidenced by powdery lesions on the leaf surfaces that ultimately cause leaf death. Aster yellows is a mycoplasma disease that causes a blanching and yellowing of the heart leaves and infected plants often bolt to produce a seedstalk that has bushy outgrowths of abnormally developed buds and sterile flowers.

Insects can also create problems in lettuce production. The cabbage looper is a 2 cm (0.8 inch) long green caterpillar that feeds on leaf tissue. The green peach aphid is not only the principal vector of lettuce mosaic virus but it also feeds on the plant and reduces its vigor and yield. This pest is difficult to control because it also colonises a number of weeds common to lettuce fields. Other aphid pests are the root aphid and lettuce seed stem aphid. The corn earworm, beet armyworm and yellow striped armyworm also are destructive insects.

Irrigation

From seedling to harvest, lettuce demands a high quantity of water. This can be provided either by sprinklers or irrigation. The usual practice is to supply early water by sprinklers and then shift to furrow irrigation when the plant reaches the rosette stage. This allows less use of water since the sprinklers can be metered to permit moisture penetration to 30 cm (12 inch) depth per setting while the lettuce plants are small. Irrigation, however, enhances the level of water related pathogen problems. Irrigated fields should be tiled so that the water can drain through the soil thus minimising waterlogging conditions.

Irrigation for lettuce seed production is generally stopped after flowering. In particular, sprinkler irrigation after flowering causes loss of seed yield since the falling water droplets dislodge the maturing seed off of the flower receptacle.

Harvesting and Threshing

A period of 12 to 21 days is required from flowering to mature seed formation. However, because of the nature of the flowering response, not all seeds mature simultaneously on the plant. As a result, lettuce plants are harvested when approximately half of the seed heads have "feathered" (seed pappus is fully expanded and dry). They can be either machine or more commonly hand harvested. If machine harvested, the plants are cut and placed in windrows until the majority of the seeds have matured and then threshed by combine. Cutting the plants by machine is best done in the morning while dew is still on the flowers. This minimises seed shattering and reductions in yield. Hand harvesting is accomplished by cutting the seed stalks and vigorously shaking the seed heads in a canvas bag or plastic bucket. This process can be repeated in two or three days to permit additional seeds to mature and increase yield.

Conditioning

Harvested lettuce seeds can be as much as one and a half times their dry weight due to contamination with high moisture flower and vegetative plant parts. These must be removed by cleaning as soon as possible to minimise seed deterioration. Lettuce seed is often cleaned on an air-screen cleaner followed by upgrading on a disk separator or an indent cylinder.

Storage

If properly stored, lettuce seeds retain their viability for four to five years. Seeds should be at a moisture content no higher than 7%. Storage deterioration of lettuce seeds is often manifested by an abnormality known as "cotyledonary necrosis". Deteriorated seeds exhibit readish necrotic spots on the cotyledons. The cause of cotyledonary necrosis is still not known.

REFERENCES

Bassett, M. J., *Breeding Vegetable Crops*, Avi Publishing, Westport, 1986.

George, R. A. T., *Vegetable Seed Production*, Longman Press, Essex, 1985.

Grogan, R. G., "Control of lettuce mosaic with virus-free seed", *Plant Disease Reporter*, 1981.

Ryder, E. J., *Leafy Salad Vegetables*, Avi Publishing, Westport, 1979.

7

Carrot Seed Production

Carrots originated in Afghanistan where they expanded to the eastern Mediterranean area. Arabs brought carrots with them during their conquests of Europe in the 10th to 12th centuries. European settlers brought carrots to the New World in the 1600s. Today, carrots are grown worldwide and its wild relative, Queen Anne's lace, is equally well known.

In early history, many of the most important medicinal plants came from the Apiaceae family of which carrot is a member. Resultantly, it is not surprising that carrot was originally used as a medicinal rather than food source. Historical writings in the 12th century extolled the virtues of carrots as a sweet root crop that could also be prepared as a juice. By the 16th century, breeding of carrots had produced a crop that showed extreme variation in size, color, and shape.

Carrots can be prepared in a variety of ways ranging from baked, boiled, steamed, and diced but their primary consumption is as a fresh product. Nutritionally, they are an excellent source of vitamin A and have moderate amounts of potassium.

PLANT DEVELOPMENT

Carrots can be grown almost anywhere as long as the growing season is relatively cool. Their optimum growth occurs between 16 to 18°C (61 to 64°F). Botanically, the carrot is an herbaceous biennial that is characterised by vegetative production in the first year of growth followed by reproduction growth after vernalisation in the second year.

Seed. The seed is fawn to brown, oblong, 3 to 4 mm (0.12 to 0.16 inches) long and 1 to 2 mm (0.04 to 0.08 inches) wide with one side flattened and the other possessing three to five ribs that are barbed. The embryo is extremely small relative to the size of the seed. The 1,000 seed weight is 0.8 g (567,000 seeds per pound).

Vegetative. Carrots form a vegetative structure consisting of an above ground rosette of leaves and a large, deep tap root. The leaves are double compound with the lower leaves of the rosette being more divided and larger than the upper leaves.

Reproductive. The carrot plant generally requires vernalisation for flowering. This effect is not related to root size since even small rooted plants, if vernalised, will bolt and flower. After vernalisation, the rosette bolts to form a stem that is 61 to 122 cm (2 to 4 feet) high with branches that produce flowering heads or umbels. The appearance of umbels is not uniform. The first and largest umbel to flower is the last umbel on the main flowering stalk and is known as the primary or king umbel. Secondary umbels are formed at the terminus of branches from the main flowering stem and flower in a sequence from the top to the bottom of the inflorescence. Tertiary umbels originate on secondary umbel stems. Figure 00.9 illustrates this flowering sequence.

Seeds produced on primary and secondary umbels vary in quality. Primary umbels produce seeds that are heavier, more mature, and of higher quality (germination percentage is higher) and vigor (germination rate is faster) than secondary umbels. As a result of this relationship, many seed companies have increased planting density in order that more primary umbels are produced for harvest. In other cases, to enhance seed yields, seed companies have established management practices that focus on harvesting the king, secondary, and tertiary umbels. This allows them to grade seeds into various sizes from large to medium to small.

Hybrid carrot seed production is possible. Male sterile lines have been discovered and in hybrid seed production the common ratio of female to male rows is 6:2 or 6:4. Unfortunately, the production of hybrid seed has required the intensive selective inbreeding of carrots for several generations resulting in inbreds with reduced plant vigor. As a consequence, yields of planted hybrid seed have not been as high as open-pollinated lines.

In addition, the male sterile flowers produce smaller petals that are less attractive to pollinating bees causing lower seed set and reductions in seed yield. These factors have led to higher costs for hybrid carrot seed. This, in conjunction with lower crop yield, have continued to make open-pollinated lines the preferred choice of planting seed among carrot growers. However, the commercial fresh market acreage in the United States of the Imperator carrot varieties is almost entirely with hybrid seed. Processing carrots are still dominated by open-pollinated varieties.

SEED PRODUCTION

Carrot production occurs throughout North America. In the winter, Florida, Texas, California, and Arizona have ideal conditions for carrot production. In the summer, carrots are grown on the muck soils of New York, Wisconsin, and Michigan. These soils are noted for producing straighter, cleaner, and higher yielding roots than inorganic soils.

Carrot seed production used to be located primarily in California. However, some years prevailed when the necessary vernalisation temperatures (below 7C/45°F) to induce flowering were not experienced or high temperatures were encountered during seed set that led to empty seeds. Today, most of the carrot seed production occurs in the Columbia Basin of Washington, Madres area of Oregon as well as Idaho which are ideal desert environments for maximum seed yield and quality.

Carrot seeds are produced by two methods: seed to seed and root to seed. In the seed to seed method, the seed is planted direct in late Summer, vernalised in the Winter, flowers initiated in the following Spring, and seeds harvested in late Summer. In the root to seed method, roots raised in beds usually in an area away from commercial production are removed from the soil in the Fall, examined for trueness to type and stored at 1C (35°F) at 95% or higher relative humidity to reduce storage rots in the Winter, and then replanted in the Spring.

The plant bed is used as a backup for direct seeded production in the event of Winter kill. Some types of carrots such as high sugar types do not overwinter well so they may be grown in a plant bed. In both cases, the process of removing plants, checking for off-types, storage, and then replanting is labor intensive and expensive. Even so, the root to seed method is required by most seed companies for new varieties, new inbreds and stock seed to allow selection for root characteristics. The seed to seed method, however, is preferred among seed companies for large commercial seed production.

Tillage

Fields are placed according to isolation distances established for each type of carrot. Carrots grow well on most soils but the best growth occurs on deep, friable soils such as sandy loams. Carrots tolerate a wide range of soil pH. Because seedling establishment is a problem in carrot seed production due to the seed's small size and slow germination rate, particular attention is give to tillage operations prior to planting. The seedbed must be clean of weeds, well pulverized, and compact to ensure ready flow of soil water to the seed and yet minimise puddling or crusting after a heavy rain.

Planting

If planting is done for the seed to seed method, row spacings of 50 to 90 cm (20 to 35 inches) are used and 2 to 3 kg of seed planted per hectare (1.8 to 2.7 pounds per acre). If the seed is mechanically precision planted, it is often coated using a ratio of 3:1 finished weight over the raw seed. Seeds should be planted 1.3 to 1.9 cm (0.5 to 0.7 inches) deep every 7 to 10 cm (3 to 4 inches). If the root to seed method is used, the transplants are placed every 20 to 30 cm (8 to 12 inches) apart in rows that are 75 to 90 cm (30 to 36 inches) wide.

Higher plant densities are desired in carrot seed production. Such a practice results in a concentration of umbels in the upper part of individual flower stalks causing a higher proportion of primary to secondary umbels and a more uniform period of seed maturation from umbel to umbel. Normally, bees are used for open pollinated varieties and are required for hybrid seed production. Usually 2 hives per hectare (5 hives per acre) are required.

Fertilisation

Because of the wide range of soil types in which carrots grow, a soil analysis provides the best recommendation for fertiliser requirements. As a general rule, the relative proportions of 1-2-2 are used. Less nitrogen is needed on muck soils compared to mineral soils. The nitrogen should be applied as ammonia since other forms cause serious burning. The fertilisers are best applied in a band and to the side of the seed at planting. Vegetative growth is further encouraged, particularly under irrigated conditions, by a second side dressing later in the Summer. In the Spring of the second year, another nitrogen application will stimulate vigorous plant growth prior to flowering.

Weed and Pest Control

Weed control in carrots is almost totally dependent on chemicals. This is because any physical alteration of the firm seedbed dislodges the small seed from the soil and results in poor emergence. In the past, herbicidal solvents such as Sotddbard's solvent at 702 to 749 liters per hectare (75 to 80 gallons per acre) were applied after the first true leaves were formed and before the four-leaf stage. This treatment delayed weed growth until the carrot's foliage was large enough to shade competitors. Another approach was to use shielded flame throwers to burn the weeds. Today, weed control in carrots is practiced using a combination of approved herbicides. Weed seeds that are particularly troublesome to remove during conditioning include canada thistle, nightshade, and barnyardgrass.

Because carrots are subject to an array of diseases and insects, crop rotation is one of the most effective production practices to minimise the buildup of these organisms. Common diseases include leaf blights, carrot yellows caused by a leaf hopper transmitted virus, and brown roots which usually infects feeder roots. Of particular note are two distinct diseases caused by the related species of the fungus Alternaria, Alternaria leaf blight and black rot. Both diseases are found wherever carrots are grown and because these fungi survive between crops in residues, both diseases cause significant yield losses where continuous cropping of carrots is practiced.

More importantly, from a seed production perspective, both fungi can be carried on or in the seed. Another seedborne disease is caused by the bacterium from Xanthomonas. As a result, most seed companies routinely use standardized seed assays to determine the level of infection of these devastating fungi and bacteria. When infected seed are commercially planted, they can cause 100% crop failure. Troublesome insects are leaf hoppers, aphids, tarnish plant bug, carrot rust fly, and carrot weevil. The major insect pest is lygus bugs which decrease seed germination. In all of these cases a comprehensive integrated pest management strategy is necessary to control lygus bugs, weeds, and fungicidal sprays for alternaria.

Irrigation

All carrot seed production is irrigated. Carrots require an abundant and well distributed water supply for optimum growth. The primary approach is to use rill (flood) irrigation. Fields irrigated overhead should be evaluated for the presence of additional fungicidal sprays. The water table should be kept at 76 to 91 cm (30 to 36 inches) below the soil surface. In inorganic soils, the water can be supplied by sprinklers and furrows. Sprinkler irrigation is most effective during the early stages of germination and emergence when the water supply does not have to be extensive. With later vegetative growth, sprinkler irrigation should be discontinued because of the increase in foliar diseases encouraged by the wet leaves.

Harvesting and Threshing

For hybrid seed production, the male rows are removed and turned into the soil prior to swathing. Harvesting of the seed crop is done by machine. This is initiated when the earliest maturing seed on the primary umbel turns brown and the umbel is brittle. Because of the uneven maturity of the seeds on the umbels and the tendency for shattering, umbels can be sprayed with a polyvinyl acetate adhesive to reduce seed loss. This process, however, is not done in large scale seed production. Harvesting early in the day while dew still remains on the umbel is also helpful. The fields are swathed by machine into windrows. After 3 to 5 days of drying, the crop is havested with modern combines at a seed moisture content of 7%.

Drying and Cleaning

The small size of the carrot seed makes it very responsive to equilibration with a low relative humidity environment. As a result, carrot seed is seldom dried unless wide ranges in seed maturity exist. Since the barbs on the seed reduce flowability and increase seed volume, they are often removed by a debearder. Cleaning of the seed is accomplished using an air-screen cleaner and indent cylinders. Carrot seed is sized into varying lengths and widths for improved precision planting.

Storage

Carrot seed should not be stored at moisture contents above 7.0%. At this moisture content and under appropriate temperature and relative humidity conditions, the seeds remain viable for 3 to 5 years.

REFERENCES

Mabey, R., *Flora Britannica,* London: Chatto and Windus, 1997.

Rose, F., *The Wild Flower Key,* London: Frederick Warne, 2006.

8

Production of Sugarbeet Seeds

Sugarbeet originated in the Mediterranean. It belongs to the same species as the red or gardent beet. In the middle of the 18th century, it was discovered that sugar from sugarbeet was identical to that found in sugarcane. From that point, countries reliant on sugarcane as their sole source of sugar begainefforts to develop this as an alternative sugar source. France was the first country that to breed a new variety of sugarbeet that could be economically processed for profitable yield. Initially, sugarbeet was introduced in 1831 into the United States in Massachusetts, but this effort was not successful. Later production of sugarbeet in California, however, did succeed and ensured that this crop remains a sable agricultural commondity in North America.

Most sugarbeet seed in the United States is produced in the Willamette Valley of Oregon and in areas of the Intermountain West, particularly Utah, Arizona and Idaho. The primary reason for this location is the favorable harvesting weather and isolation from other sugarbeets which might be a source of genetic outcrossing and contamination. The seed is produced by specialised seed companies associated with different sugar companies under contract with seed growers by methods and standards prescribed by the seed company.

PLANT DEVELOPMENT

Sugarbeet is a biennial plant that grows vegetatively the first year and produces seed the second year after being vernalised by thermal induction during the Winter. However, when grown as a sugar crop, it is harvested after the first year during which the leaves and the fleshy root are formed. In the first year, it produces a dense canopy of leaves and a large root in which the sugar is stored. In the second year, the sugar is used to form the flowering stalk, flowers, and seed. The leaves are simple with a large blade and petiole. The root is divided into three zones: The top from which the leaves are produced in a dense spiral, the thick fleshy hypocotyl where most of the sugar is found, and the root which becomes a long taproot. The flowers are perfect but incomplete; they have a five part calyx but the corolla is absent. They are produced on a panicle-like spike.

Traditionally, seed was produced by digging up the roots after the first year's growth and storing them in cool, damp storage over the Winter. These propagules, called stecklings, were replanted the following spring to produce seed. However, it was discovered that thermal induction could also be obtained by growing the crop as a Winter annual. By planting in the late Summer or early Fall, seed could be produced the following Summer. This method is used today in North America and most areas which produce sugarbeet seed.

SEED PRODUCTION

The sugarbeet "seed" that is planted is actually a single achene, which is a one-seeded fruit with a somewhat inflated, corky outer layer, or pericarp. The traditional natural sugarbeet seed unit was a multi-germ "seedball," or a cluster of two to five achenes formed by the joining of several floral units. When planted, each of the achenes germinated, resulting in three to five plants. The resulting stand of plants was much too thick for optimum root and sugar production and had to be manually thinned at great expense to the grower.

In 1948, a single plant was identified which produced single monogerm seeds or a single achene. From this and other sources, new monogerm sugarbeet varieties were rapidly developed and by early 1960 had almost completely replaced multigerm seed. By 1960, open pollinated varieties had also been largely replaced by single cross, three-way cross or double-cross hybrids.

It is important that sugarbeet seed producers have a contract with seed companies or sugar companies for seed production. This assures access to varieties that will be in demand by growers. It also assures them of parent seedstocks (i.e., inbred lines) needed to produce seed of different hybrids. Furthermore, it insures seed growers of technical advice needed for managing the planting of parental and restorer lines. Most importantly, it assures them of a market for their harvested seed.

Hybrid seed is produced in two different ways. Most commercial seed is produced by mixing the male-sterile F1 line as the primary female parent with seed of a multi-germ pollinator in a 10:1 proportion. The resulting crop contains a low percentage of seed that is not hybrid. However, since these are large and shaped differently, they are easily separated from the small flattish monogerm seed during conditioning.

The second method is more expensive, but results in a completely hybrid crop. It is used primarily to perpetuate parent stock seed which will be used in three-way or double-cross hybrids. It is produced by growing the multi-germ pollinator in about four separate rows separated by a skip-row from 16 to 20 female male-sterile rows. The male rows are destroyed after pollination and before harvesting the hybrid seed produced on the male-sterile parent.

The male-sterile lines are perpetuated for further use by backcrossing with their type "O" equivalent. This process insures that the progeny will be male sterile. The type O lines are male fertile lines that do not restore pollen production to the offspring when crossed with male-sterile lines. They are comparable to maintainer lines used in producing hybrid corn.

Tillage

Sugarbeet grows best on slightly acidic fertile loamy soils. Sandy soils do not retain adequate moisture between irrigations and heavy clay soils cause problems at harvesting. Fields are plowed to 20 to 30 cm (8 to 12 inches) deep in the late Summer. If sprinklers are not used for irrigation, the fields are leveled for surface irrigation and a firm seed bed prepared by disking, harrowing, and rolling to minimise soil crusting.

Planting

Planting occurs in late August or early September. Seed is planted with a standard drill in rows spaced 51 to 60 cm (20 to 24 inches) apart at a rate of 13.4 to 16.8 kg per hectare (12 to 15 pounds per acre). Seeds are platned from 2.5 to 3.8 cm (1.0 to 1.5 inches deep. The optimum temperature for germination is 15°C (60°F). Temperatures below -4°C (25°F) kill beet seedlings.

The fields must be isolated from other fields of sugarbeets or garden beets that are allowed to produce seed. This assures genetic purity of the resulting seed. For varieties that are similar in characteristics, isolation of 0.8 km (0.5 miles) may be adequate. However, for elite seedstocks or very dissimilar varieties, isolation of 3.2 to 4.8 km (2 to 3 miles) may be desirable. It is important that the guidelines of the seed company for each variety be closely followed.

Fertilisation

Fertility practices vary greatly between seed production

areas. In Oregon, the usual practice is to apply 268 kg per hectare (240 pounds per acre) of nitrogen and 112 kg per hectare (100 pounds per acre) of phosphate, along with 56 kg per hectare (50 pounds per acre) of sulfur and 5.6 kg per hectare (5 pounds per acre) of boron. Another 179 kg per hectare (160 pounds per acre) of nitrogen are side-dressed in the Spring, often in two applications.

In Arizona about 56 kg per hectare (50 pounds per acre) of nitrogen are applied prior to planting, along with 67 kg per hectare (60 pounds per acre) of phosphate. Another 112 to 168 kg per hectare (100 to 150 pounds per acre) of nitrogen are added during the Fall and an additional 224 to 336 kg per hectare (200 to 300 pounds per acre) are side-dressed in the early Spring or added with the irrigation water. In Utah 269 kg per hectare (240 pounds per acre) of phosphate and 140 kg per hectare (125 pounds per acre) of phosphate are applied prior to planting and another 67 kg per hectare (60 pounds per acre) of nitrogen are side-dressed in the Spring.

Weed and Pest Control

Good weed control includes cultivation and chemical weed practices in addition to hand weeding. Sugarbeet grows rapidly and soon shades most weeds and prevents them from establishment. Cultivation should be used wherever possible. However, chemical control is the most common method and many chemicals are registered for all seed production areas. Finally, hand weeding is an important additional weed control practice. The most common weeds of sugarbeet seed fields are pigweed, lambsquarter, barnyardgrass, mustard, and wild oats.

Rarely can sugarbeet seed crops be grown continuously on one field. Rather, they are often rotated with cereals, beans, vegetables such as tomatoes and lettuce, and forage legumes to control diseases and root-knot nematodes. Black root is a fungal disease that kills germinating seedlings and reduces plant vigor so that uneven stands and poor yields result. Seed treatment is effective against this disease. The virus disease, curly top, is a particular problem for sugarbeet seed producers in the intermountain area. This disease is controlled almost exclusively through the use of disease resistant varieties. Virus yellows is another disease which occurs throughout the West that may also be destructive to both yield and quality of seed.

These diseases are spread by leafhoppers or aphids. Although neither disease is seedborne, both can be destructive and reduce seed yield. A variety of insects attack the sugarbeets including beet armyworms, beet webworms, beet leafhoppers, and grasshoppers. Most losses in seed yield are attributed to these insects feeding on the foliage of mature plants. The root-knot nematode is a microscopic eelworm that persists in the soil, feeds on the root, and can cause serious losses in seed yield. Control is accomplished by soil fumigations or crop rotation. Seed producers will make 5 to 6 applications of chemicals to control weeds, diseases, and insects throughout the Fall and Spring prior to harvest.

Harvesting

Sugarbeet seed shatters readily as it approaches maturity. Shattering is minimised by windrowing the crop before the seed reaches full maturity and allowing it to cure a few days before combining. Windrowers are equipped with vertical cutter bars on each side to enable the crop to be cut with a minimum of shattering. In drier areas such as Arizona and Utah, this requires about 7 to 10 days compared to about 10 to 14 days in the Willamette Valley. The cutter-bar should be raised high enough so the plants are gently laid on the ground so air can circulate through the windrow and allow aeration and drying. Some of the mature seed will still shatter during windrowing, however most will fall onto the windrow and can be recovered during combining.

Sugarbeet plants tend to be large and viny, thus the size of the windrow feeding into the combine is substantial. However, this should not cause a problem if the crop has been adequately dried and the combine properly adjusted. Each windrow may consist of three to six rows, depending on the size of combine used for threshing. The direction of the combine should be the same as that of the windrowing operation because the seedheads are laid down to go into the cylinder head first.

Some of the small, lighter seed will be lost from the back of the combine with the tailings. These fall to the ground and provide a source of volunteer plants in subsequent crops which can be eliminated by light tillage which encourages their germination so they can be destroyed by further tillage.

Conditioning

The harvested seed is first conditioned over an air-screen cleanter to remove inert material as well as multi-germ seeds which contaminate the lot. This is followed by a draper belt which further removes stems and remaining inert matter. The remaining multi-germ seed is removed by width or thickness graders which allow the small monogerm seed to fall through, but reject the large multi-germ seed. The monogerm seed are further conditioned by passing between two horizontal disks, one of which rotates. This action rubs the corky tissue from the seeds and reduces them to small, uniform size and shapes. Finally, the seeds are conditioned over a gravity table which removes light weight, low quality seed from the seedlot. The seed is then treated with a fungicide and an insecticide.

REFERENCES

Cooke, D. A., and R. K. Scott (eds.), *The Sugar Beet Crop*, Chapman and Hall, New York, 1993.

Johnson, R. T., J. T. Alexander, G. E. Rush, and G. R. Hawkes (eds.), *Advances in Sugar Beet Production*, Iowa State University Press, Ames, 1971.

9

Bean Seed Production

The common bean has the largest number of varieties, and the most culinary uses. The various varieties of common beans are well known for their ability to produce seed coats in a wide array of colours covering nearly the whole colour spectrum. In a few cases, the seed coat colour is distinctive enough to identify a particular variety of bean, though the colour darkens with age. Varieties of beans can be differentiated not only by seed colour and size, but also the size, colour, and shape of their leaves, the degree of pubescence (plant hairs) on the pods and leaves, and the colour of their flowers.

Beans are herbaceous annuals with some perennial tendencies, such as the slightly woody stem found in vining types. Depending on variety, plant height ranges from dwarf (bush plants) to semivining (half-runner plants) to vining (pole) beans. Vining varieties have the ability to twine or climb a pole or support, but the degree of twining may vary slightly according to variety. All beans have pinnate, three-lobed leaves. They are usually broad ovate in shape, but certain varieties may have long narrow leaves.

PLANT CHARACTERISTICS

The lima bean (named for Lima, Peru) has a flat broad seed ranging in size from ½" to 1" or larger. Like the common bean, the lima also has a number of colour and size variations. There are two subclasses of limas: the small-seeded lima (such as the sieva lima), and the potato lima, which has a thicker seed.

Runner beans are less variable in size and colour. Typically, the seeds are various shades of maroon with mottled light-purple streaks, though some may be white or pale green. Unlike other beans, the cotyledons remain below ground when the seed germinates. White-seeded runner beans have white flowers, while coloured-seeded runners have scarlet-coloured flowers. Runner beans, native to Central America, are perennials that form tuberous roots, but in northern climates they are cultivated as annuals.

Though beans are predominately self-pollinated, various varieties of beans differ in their ability to cross-fertilize. Cross-fertilisation does not occur between different species of beans (for example, between a common bean and a lima bean). For that reason, each species may be grown side-by-side without concern for crossing. On the other hand, when growing varieties of the same species, varieties need some degree of isolation from each other in order to maintain purity.

Climate and Regional Growing Requirements

Successful bean seed production is more dependent on climate than most other vegetable seed crops. Several factors are important here. In hot humid regions, special harvesting and drying techniques are required to produce high-quality seed. Also two seedborne diseases, anthracnose and bacterial blight, can be a problem in humid, wet climates. That is why much commercial bean seed production has moved to dry western states.

Beans are considered a warm-season crop, sensitive to temperature extremes. Several days of extended hot weather may cause failure of fertilisation and subsequent blossom drop, resulting in lower yields. Cool, rainy weather also has an adverse effect on yield. Lima beans tolerate hot weather better than common beans, and can be grown further south, though some lima varieties may not set seed well when the temperature exceeds 95°F (35°C). On the other hand, most limas do not tolerate cool growing seasons as well as common beans.

Runner beans, which are indigenous to the cool, humid uplands of Guatemala and southern Mexico will not set seed well in the Mid-Atlantic except at higher altitudes, or during the cooler days of early summer or late summer/ early fall. Beans are fairly tolerant of a wide range of soil types, though have a more limited tolerance of acid or alkaline soil than other vegetables. In the Southeast, where soils are more acidic, proper attention to pH can be helpful for good production. Soil pH should be in the range of 5.8 to 6.5.

Determinate and Indeterminate Varieties

Determinate (bush) varieties have short stems with four to eight nodes on the main stem with short internodes. Semi-determinate (half-runner) varieties have stems two to four feet in length. Indeterminate (pole) varieties grow up to ten feet in length with a twining growth habit, and little branching of the main stem. The internodes of indeterminate varieties are longer and only the length of the growing season limits the number of nodes produced.

Roots

Bean plants develop taproots, that in good soil can reach a depth of at least two feet within a month, and three to four feet when fully mature. Strong lateral roots also develop, and a fully mature vining variety may have a root zone radius of two feet. Nitrogen-fixing nodules may or may not be present on the roots.

Flowers

Beans have perfect flowers (male and female reproductive parts are in the same flower). The flowers are typically white in many of the newer varieties of bush beans, but flower colour, especially in older varieties and vining varieties, may be white, pink, rose, lavender, deep violet, or sometimes yellow. In runner beans, they may be crimson, or crimson and white. The flower consists of a corolla (united petals) and a keel (the two lowest petals joined at the lower edges, resembling the keel of a boat). Usually there are ten stamens, the upper one free, while the other nine unite into a twisted tube that encloses part of the style and the long ovary. The flowers are borne along a two to three-inch rachis (stem).

Pollination and Fertilisation

Pollination is the process of transferring the pollen from the male part of the flower (anthers) to the female part, or stigma (pollen-receptive surface of the pistil). Fertilisation is the process of the union of the male gamete in the pollen with the egg to form the zygote, which becomes part of the seed.

Beans are more susceptible to crossing than are peas, and certain bean cultivars are more susceptible to crossing than others. Beans are primarily self-pollinated and self-fertilised, but natural crossing often occurs as a result of pollinator pressure, principally by large bees, such as the bumblebee, and to a lesser extent by the honeybee.

Blooms open in the morning, between 7 and 8 a.m. and never close, though the corolla is shed within a few days of blooming. The stigma is receptive two days before the flower opens and even though the pollen is shed from the anthers the evening before the flower opens, the pollen isn't transferred to the stigma at this time unless the flower is agitated or tripped by a bee. After the flower opens, the flower can be cross-pollinated by bees. A short time after opening, the stigma retracts into the keel and comes in contact with pollen from the anthers and self-pollination occurs. After pollination, the pollen tube grows toward the ovary and fertilisation takes place about 8 to 9 hours later. Pollen tubes from foreign pollen may grow more quickly than self-pollen, leading to a higher incidence of cross-fertilisation.

Common beans and limas seem to set a full complement of seeds and pods in the absence of bees, and most studies indicate that bee pollinators do not have a significant effect on yield. Though bees are not essential for seed development and increased yields, they can be a cause of crossing if the flowers are pollinated early in development. On the other hand, runner bean yields can be increased significantly as a result of insect visitation.

Seed and Pods (fruit)

The fruit of the bean plant is a bi-valved pod that bears seeds alternately on both sides of the dorsal margin of the placentae. The characteristics of pods differ among varieties: they vary in size, shape, length, and curvature of the pod as well as the width of the cross section, which may be round or flattened. Also varieties differ in the amount of seed within the pod and how well the pod dehisces (splits open) along the dorsal and ventral side.

Most cultivated bean varieties have to be threshed to remove the seeds from the pods, but very old or ethnic varieties have a tendency for the pods to shatter, causing the seeds to be flung out by the pods twisting into a spiral "spring" as they dry.

Seed Germination

The statistics on seed germination in relation to temperature are given in the table below:

Type of Bean	Minimum (°F)	Optimum Range (°F)	Optimum (°F)	Maximum (°F)
Common bean	60	60 to 85	80	95
Lima bean	60	65 to 85	85	85
Runner bean	60	60 to 85	80	85

In the optimum temperature range, common beans will emerge from the soil in about 6 to 15 days. Lima beans will emerge in about 7 to 20 days, depending on temperature. Though 60oF (16°C) is considered the minimum soil temperature for germination, significantly better emergence occurs when the soil temperature at 8 a.m. is at least 65°F (18°C). When the soil is cool, seeds may rot before they germinate. For early spring planting, dark-seeded bean varieties are recommended because they have a better germination success in cool soil than do light-seeded beans. This is because the pigments in the seed may provide some protection from opportunistic fungi.

ISOLATION DISTANCES

Though bean flowers are perfect and self-pollinating, a considerable amount of out-crossing can occur under certain conditions, and therefore isolation between varieties is required. The important point for the seed grower to remember is that isolation distance recommendations need to be understood within the context of the environment in which the crops are grown. This is especially important for organic growers who, as a result of their growing practices, have a larger degree of biodiversity in the environment, and therefore a higher pollination pressure on seed crops.

Some of the variables that influence the degree of outcrossing are the following:

— variety characteristics such as flower structure,
— environmental factors such as wind and temperature,
— types of pollinators and their behavior on the blossoms,
— isolation distance,
— presence of barrier plants,
— planting patterns such as row or block plantings,
— number of varieties,
— number of plants of each variety,
— presence and type of other pollen sources, and
— regional or bioclimatic factors.

Because beans have such distinctive colours and patterns in the seed coat, it is tempting to conclude that crossing has occurred when a change in the seed coat is observed. While this may be true in some instances, the absence of such a change does not constitute evidence of seed purity. Following are some data summarised from pollination literature regarding the amount of outcrossing in beans planted in adjacent rows: (1) common bean (Phaseolus vulgaris) - 1 to 25%; (2) lima bean (Phaseolus lunatus) - 1 to 89%; and (3) runner bean (Phaseolus coccineus) - at least 32% cross-pollination reported in one study, but runner beans cross more readily than lima beans and cross-pollination may exceed 90%.

The extent of crossing is highly dependent on the microclimate and biodiversity of the region in which the beans are grown for seed. As a beginning seed grower it is important to be very conservative in establishing minimum isolation distances until an understanding has been gained about the dynamics that can affect crossing in beans.

Isolation distances are determined partly on the intended use of the seed. The chart below shows minimum recommended isolation distances for different types of beans. These distances can be modified on the basis of variables such as the presence of other pollen sources throughout the flowering season, or the presence of tall, dense, physical barriers. If you are a beekeeper and a seed grower, it is best to grow only one variety of lima bean or runner bean because bees can easily forage ¼ mile or significantly further.

Under extremely high pollinator pressure, lack of other pollen sources, and no barriers, the isolation distances for lima beans and runner beans may need to be increased if you are growing "registered" or "elite" seed. Less separation is required if seed is collected only from plants in the center of block plantings. A tall barrier crop such as ornamental sunflowers is recommended.

Borders of perennial flowers are very useful for attracting bees, especially those flowers that have open, exposed nectaries, for example, members of the Composite Family (such as daisies and coneflowers). If necessary, bush beans can be caged to exclude pollinators. The simplest solution is to build a spun-polyester hoop house. Though there is an initial expense, the components for cages can be reused year-after-year, except for the polyester which may need to be replaced every two or three years. A hoop house can be constructed by inserting one-foot lengths of rebar, electrical conduit pipe, or PVC (½" inside diameter) pipe into the ground at two-foot intervals along both sides of a double row of beans.

To make the hoops, purchase flexible black plastic water-supply pipe (¾" inside diameter). This is available in 100-foot rolls at building supply stores. Cut the pipe into sections about five or six feet long, bend into a hoop, and then insert the loose ends over the upright metal bars. The hoops are then covered with spun polyester. Edges of the polyester can be held down by boards, soil clips, or buried in the soil. The hoop house not only excludes pollinators, but also may exclude Mexican bean beetles if constructed early in the season (and if set up where Mexican bean beetles are not likely to emerge from the soil). One caution regarding this technique: beans grown under spun-polyester row cover may be subject to higher temperatures which may have an effect on seed set.

Evidence of Crossing

If crossing has occurred within a flower, the results of the crossing will not be visible in the seed derived from that flower. The seed coat colour and pattern will be the same as the parent seed. Evidence of crossing will not be visible until the second generation, when evidence of crossing can be seen as individual variation within the population of plants, and the seed collected from those plants. Changes in seed coat colour are a good indicator that crossing has occurred, but not all crosses show up as a change in the seed coat colour. Therefore, stability of seed coat colour is not a reliable indicator of seed purity.

If you have discovered that crossing has occurred, both the seed from your current crop and the seed used to produce that crop is impure and unsaleable. Therefore you will need to discard the seed from both your current crop and the previous crop, and start over with pure seed from a previous pure generation.

To understand the origin of seed coat colour, it is important to know what tissue is responsible for seed colour. In beans, the seed consists of both embryo and maternal tissue. The embryo is the result of either self- or cross-fertilisation, and therefore is the next generation. So if crossing has occurred, it will be expressed in the embryo, not the seed coat, which is maternal tissue. The embryo has two large cotyledons which are the food storage organ of the seed.

Though the cotyledons of the embryo can have a distinctive colour characteristic of the variety, the cotyledon colour is not usually visible underneath the seed coat of the dry seed (though conceivably, it could in a rare instance, affect seed coat colour). It is the testa (seed coat) which is derived from maternal tissue that gives rise to the seed colour. Therefore you can't determine by looking at the seed, whether a cross has occurred, until you grow out the seed and observe its progeny.

Common beans and lima beans are naturally inbreeding plants. Runner beans are not so easily classified, though perhaps they can be considered naturally inbreeding, with outbreeding tendencies. For all intents and purposes common, lima, and runner beans can be treated as "selfers" and therefore not subject to inbreeding depression, characterised by a loss of vigor and genetic variability.

Many modern commercial varieties are highly uniform and have little heterozygosity relative to some heirloom varieties which can be somewhat variable. An example of a non-uniform variety is the heirloom pole bean, 'Turkey Craw', which appears to be uniform with regard to foliage and seed coat colour pattern, but has two slightly different types of pods.

The difference in the pods can be seen when they are dried: one is more wrinkled than the other and there appears to be a slight difference in threshing characteristics in that the wrinkled pod retains the seed more easily. In theory, as many as 190 to 200 plants are required to maintain the genetic diversity of most selfers. The greater the tendency for natural outcrossing, the larger the population size needed to maintain genetic diversity within a variety. Thus for runner beans, the minimum population size should be higher, perhaps 40 or more plants. For some insight on this, it is useful to refer back to the section on isolation distances where it is noted that outcrossing tendency of runner beans is greater than that of limas which is greater than that of common beans.

SELECTION AND ROGUING

One form of selection that can be practiced by the grower is roguing. Roguing is the process of removing off-type plants. Varieties are never static and are always changing. For that reason there is no such thing as "variety maintenance". Selection is always happening, whether we have a hand in it or not. There are environmental factors (climate, weather, soil, diseases, pests, occasional crosses, etc.) and genetic factors (mutations, genetic drift due to population size) which are continually changing the variety. Thus you have to practice roguing to maintain desirable variety characteristics. Roguing is done throughout the growing season, from seedling stage through fruiting stage, whenever the need occurs. If possible, plants should be rogued before flowering so that the undesirable traits don't pass into the seed.

Roguing is done by inspecting the entire plant, not just the fruit or seed. When roguing plants, pay special attention to earliness, foliage colour, leaf shape, flower colour, growth form, trueness-to-type, vigor, and disease and insect resistance. When pod production begins consider for example, productivity, size, colour, shape, and eating quality. Your focus should be on the whole plant, not on individual pods. Finally, roguing can be done again on the seed itself. If you find off-types at this stage you may have to evaluate the purity of your seed crop.

CULTURAL CHARCTERISTICS

Soil Preparation

Successful bean production depends on careful preparation of the seedbed. Once the seedbed has been prepared, it is helpful to pre-irrigate (which is preferable to irrigating after planting) so that the soil is uniformly moist. If the soil is not pre-irrigated, it is important to lightly firm the soil over the beans in the trench so that the soil is in close contact with the seed, otherwise germination may be slower and more erratic. If the soil is not moist, the crop should be irrigated immediately after planting. If the soil is prone to crusting, it may be necessary to irrigate again to help the seedlings emerge through the heavy soil.

Beans do best in a soil which has good aeration and drainage, such as a sandy or clay loam, but do not thrive well in heavy, cold, clay soil. The soil should be well worked and supplemented with organic matter to improve soil tilth, nutrition, and to encourage beneficial microorganisms. Green manure should be turned under at least four weeks before planting. Soil pH should be in the range of 5.8 to 6.5.

Sowing

Common (bush and half-runner) beans: Beans are very frost tender and need a minimum average soil temperature of 65°F (18°C) to germinate well. Sow after the last frost-free date. Dark-seeded bean varieties resist rotting in cool soil, and can be planted when the soil temperature is as low as 60°F (16°C). Plant seeds ¾" to 1" deep, spaced 2 to 3" apart within the row and thinned to 3" apart. Bush varieties can be spaced in rows 18" apart. Though close spacing helps suppress weeds, too close spacing makes it more difficult to work the crop and to scout for bean beetles, and therefore minimum row spacing of 20 to 24" is recommended for most bush varieties. Semi-runners will need wider row spacing, at least 36" apart depending on the variety. The sowing rate for bush and half-runner beans ranges from 45 to 95 pounds of seed per acre depending on the variety and location.

Common (pole) beans: Pole beans bear later than bush beans and give higher yields. Pole beans grown commercially for seed in western states are grown on the ground without support, but in the Mid-Atlantic and South they should be grown on poles or trellised to reduce disease and insect problems. Poles should be at least eight feet high, and 36" apart. If poles are used, they should have a rough surface so that the vines can climb without sliding down the pole. Sow six to seven seeds, ¾" to 1" deep, about 6" out from the base of each pole. Pole beans can also be grown on tripods made of poles eight feet long. These are spaced about four feet apart center to center.

Trellising is another option. Trellises are constructed by placing metal or wood posts eight feet apart with heavy-gauge wire strung between the tops of the posts. Trellis material is then suspended from the wire. If trellis material isn't available, a twine trellis can be constructed by stretching twine between the posts about 6" above the ground. Twine is then woven, zigzag fashion between the top wire and the bottom wire. To prevent slippage of the twine toward the center, the twine must be lashed to the wire and bottom twine. Alternatively, the twine can be tied to the top wire and hung straight down and then tied to the bottom horizontal twine. The sowing rate for pole beans ranges from 20 to 40 pounds of seed per acre depending on the variety, location, and cultural method.

Lima beans: Lima beans need a growing season of about 3-1/2 to 4 months. Culture is similar to common beans, but the root system is more extensive and the plants need a wider spacing. Lima bean seed should not be planted until the soil temperature is at least 70°F (21°C) and preferably

75°F (24°C), or typically two to four weeks later than common beans. Avoid planting limas in soggy soil. Irrigating seeds after planting can cause considerable rotting of seeds. Highest yields of limas are obtained when spaced at 3" apart in the row.

When sowing pole limas, sow five to six seeds at the base of each pole, about 6" out from the base of the pole. The sowing rate for lima beans depends on the variety, the thickness of the stand desired, the size of the seed, the location, and the culture method. Average sowing rates are as follows: bush limas, 40 to 60 pounds per acre; small-seeded bush limas, 60 to 90 pounds per acre; and pole limas, 30 to 40 pounds per acre.

Runner beans: Runner beans (both bush and pole types) are grown in a manner similar to limas, except they can be started earlier in the season. The sowing rate typically averages 60 pounds of seed per acre, but like limas, variety, location, growth habit, and culture method all affect sowing rate.

Cultivation

Areas between bean rows can be weeded and tilled fairly deeply when the plants are several inches tall. Later tilling of the soil should be shallow to avoid disturbing the roots, and once the pods have formed, hoeing or tilling is not recommended, nor should it be necessary. A loose layer of straw mulch between the rows encourages predators such as spiders, and helps retain soil moisture. If the crop is irrigated from above, it should be done early enough in the day so that foliage is dry by evening, otherwise moisture on the leaf surface may encourage bean rust.

HARVESTING

Bean seeds are ready to harvest as soon as the pods are brown and dry. Most commercial bean seed production is done in the dry western states where pods can be left on the plant to dry until harvest, without fear of disease. Bean seed growers in the Mid-Atlantic and South are not so fortunate and have to pay special attention to the weather and the timing of harvest. Weather permitting, the pods are left on the plant to dry, but most growing seasons the weather is too humid or rainy to allow the pods to remain on the plants.

Extended wet weather may cause the pods to become moldy, soggy, and useless and beans may sprout within the soggy pods. Any seed that is salvaged under these conditions will often be discoloured, moldy, and of low viability. You might as well throw the seed out. If you are fortunate to have dry weather, the pods can be left on the plant to dry, but if wet weather threatens the pods must be harvested before the rain. A single thunderstorm may not be a problem, but several consecutive thunderstorms over several days may be enough to cause problems with the seed. During this stage the pod should be somewhat flaccid with no more than 10 to 15% green colour present.

The partially dry pods are then brought under cover or indoors to complete the drying process where conditions can be controlled. The pods are spread out in a thin layer about 1 to 2" deep on tarps, sheets, screens, trays, or other suitable surfaces. At this point, the goal is to dry the pods as quickly as possible. Ideally, this should be done indoors in an air-conditioned environment, but this is not always feasible. If not possible to dry indoors, a shed or garage with good ventilation is another possibility. If a fan can be used to ventilate the space, all the better. The pods may need to be turned periodically. Once dry, the pods should cure for another two to three weeks. If drying in a barn or shed you will need to watch for rodents and take whatever preventative or control measures necessary.

At the end of the growing season, if frost threatens the crop, the entire plants can be pulled and hung upside down in a warm area so that the pods continue to dry. This allows the beans and pods to mature for several more days. Not all the pods will be salvageable, but a considerable portion of the harvest may be saved.

THRESHING AND WINNOWING

Beans should be fully cured before being threshed. During the curing process the pod appears dry, but the seed continues to lose water. The seed gradually changes from a rubbery consistency (when it can be bent and dented with a fingernail) to a hard object that can be shattered or broken. Though beans should be hard before being threshed, they are, nevertheless, a live organism subject to damage.

Threshing is the process of using a mechanical force to release the seed from the pods. Threshing is a skilled art, whether done by machine or done by hand. Seed which has been subject to improper threshing may appear intact and normal, but may have small cracks or internal damage that produces seedlings with developmental disorders that ultimately lead to seedling death or stunted growth. Properly cured seed will be seed that has a moisture content high enough to prevent small cracks upon threshing, but low enough to prevent internal damage to the embryo.

Methods for Threshing Bean Seed

Following are some threshing methods that can be used for threshing several pounds to a hundred or more pounds of beans. There are a number of possibilities for threshing. The method of choice is related to the volume of seeds to thresh and the equipment available. Also choose a dry day to do the threshing: the lower humidity will increase the efficiency of the process.

Feed Sack Method

The dried pods are placed in a large feed or grain sack and the opening is tied shut with a piece of twine. The sack is then placed on the floor or a cement slab where you can jog in place, twist, or dance on top of the sack. Turn the sack and shake the contents occasionally to redistribute the contents. Soft-soled shoes should be used for this because stomping on the seeds with hard-soled shoes could damage the seed. The idea is to apply just enough pressure to separate the seeds from the pods. You can also vary the routine by swinging the sack against a brick wall or down on a driveway. After most of the seed is threshed you can massage and twist the sack with your hands to break up the pods further.

Canvas Ground Cloth Method

A canvas ground cloth or plastic tarp can be used, but canvas is more durable. You can find a large canvas tarp in the painting section of many large home improvement stores. Simply dump the pods onto the tarp and jog around on them until the pods are broken up. Alternatively, you can beat the pods with a flail, but jogging on the pods might cause less seed damage.

Hand Threshing Method

Threshing by hand is a good method for processing small amounts of seed. The pods can be either split by hand or placed in a plastic tub and wrung through your hands, in which case you'll want to wear gloves. Methods for cleaning and winnowing seed:

Screening

Most of the chaff can be removed from the seed by screening. Screens can be constructed from hardware cloth which comes in three-foot wide rolls available at hardware and home-improvement stores. Hardware cloth is available in ½", ¼", 3/8", and 1/8" size mesh. The latter two sizes may be harder to find, but may be specially ordered. Hardware cloth can be mounted on a square made of furring strip, 12" on the side, which is the standard size for small seed-cleaning screens.

For processing large batches of seed, hardware cloth can also be mounted on three-foot square, 2 x 4" frames. Beans and chaff are poured through the screens. Large chaff is left behind on the large-mesh screen and small chaff will pass through the small-mesh screen leaving the beans on the screen. Start with the ½" mesh screen to remove the rough chaff, and then progress to a smaller mesh size, as necessary. The 1/8" mesh screen is used to remove the "fines" (the small chaff), though winnowing will serve the same purpose.

Winnowing:

Winnowing is the process of removing chaff from the seed, chaff that is often the same size as the seed. It can be done several different ways. Regardless of the method used, it is often useful to spread a large tarp under the winnowing area so that seed accidentally blown out with the chaff can be recovered.

— *Vacuum cleaner hose discharge method*: Some small hand-held vacuum cleaners (Hoover, for example) allow the user to mount the intake hose on the discharge side of the vacuum. This allows the user to direct a strong stream of air toward a large container of seed and chaff. A large stainless steel mixing bowl, 16" in diameter works well. This size bowl is deep and wide enough to hold several pounds of seed and the sides shallow enough to allow chaff to blow over the edges when air is directed over the seeds. It helps to stir the seeds with one hand while directing the hose with the other hand. While stirring the seed, the lighter chaff will float to the top and will be blown out by the air stream.

— *Box fan method*: Seed can be winnowed by pouring seed from one container into another in front of a variable speed fan. A variation on this method is to spread a tarp in front of the fan and pour the seed slowly and directly onto the tarp. The fan will need to be elevated a couple of inches above the tarp. The light chaff will blow off the far end of the tarp and the seed and heavier chaff will blow to separate zones at different distances from the fan.

— *I-Tech Seed Winnower*: This device is ideal for small-scale seed cleaning but is no longer being manufactured commercially. Fortunately the University of California at Davis maintains a Web site that offers plans for constructing seed winnowers and other seed processing equipment. The electric seed winnower is the easiest to make: the only materials that are needed are plastic pipe and a squirrel cage blower.

— *Clipper Office Tester and Cleaner*: This desktop-size machine is used by seed companies, seed growers, seed stores, and university labs for cleaning, sizing, and grading small amounts of seed.

— *Final inspection and hand-picking*: The last step in seed processing is to inspect the seed and to hand-pick any seed which is misshapen, stained by fungus, split, cracked, partly sprouted, or off-type. For a good batch of seed this step goes very quickly, but when seed is grown under wet conditions, considerable hand-picking is needed to remove stained seed. If the seed has been threshed too roughly there will be a number of seed with splits and cracks.

Seed Yield

Yield is dependent on variety, density of the planting, fertility of the soil, microclimate, weather, and other factors. An acre of beans raised for seed yields on average about 1,300 to 1,500 pounds of seed. A very good yield is 2,000 pounds. An average yield for an acre of lima seed is 1,950 pounds, up to 2,500 pounds for a very good yield.

Seed size and weight varies depending on variety and type of bean. An ounce of common beans contains 100 to 125 seeds per ounce, and limas 25 to 75 seeds per ounce. The Federal germination standard for commercially grown common beans and limas is 70%, and 75% for runner beans.

Labeling

It is easy to lose a good batch of seed if you have not labeled your seed at every step from extraction through drying. Never assume that you will remember which variety is which. The label should travel with every step of the process from start to finish.

SEED TREATMENT FOR BEAN WEEVILS

Seed that is not threshed and cleaned in a timely fashion may be ruined by bean weevils (Bruchus obtectus). Seed should be cleaned and threshed as soon as it is ready because it is easier to detect weevil damage in cleaned seed than when the seed is still in the pod. Weevils can be killed by freezing the seed for three days. Because the bean weevil is so prevalent in different areas of the country, this should be a routine precaution. Once seeds have been frozen they should be stored in airtight containers or in an environment free of weevils.

SEED STORAGE

Long-term Storage

It is very important that seed be thoroughly dry. Freshly harvested seed should be cured at least two weeks before threshing, and at least three to four weeks before placing in long-term storage. At least one week before the seeds are placed in airtight containers, they should be transferred to an airconditioned or heated room or other low humidity environment before being placed in a jar with a rubber-gasketed lid. Thereafter the seed may be transferred to the refrigerator or other cool environment. Before opening the jar, always allow the jar to warm up to room temperature. Try to do this on a dry day so that humid air is not introduced to the jar. Note: never store seed in a closed container unless the seed has been dried thoroughly first.

Short-term Storage

For short-term seed storage, it is not necessary to store seed in airtight jars. Seed can be stored in large envelopes, paper bags, cloth bags, or other non-airtight containers. Porous containers are not recommended for storing seed for long term unless the surrounding air remains cool and dry. Zip lock bags and most plastics allow water vapor to pass through and therefore these materials are for short-term seed storage only.

SHIPPING SEED

Just the basic information is included here. Before seeds are shipped, there should be labels on both the inside and outside of the seed containers. Paper bags should be triple bagged. If a seed container breaks open and mixes with other seed, the seed will be worthless. Any shipment of seeds should be able to withstand a minimum ten-foot drop without damage to the container or its contents.

INSECT PESTS

Mexican Bean Beetle (*Epilachna varivestis*):

The Mexican bean beetle (MBB) is one of the most troublesome insect pests of beans. MBB is easily controlled by a small parasitic wasp (Pediobius foveolatus) native to India, now raised in insectaries for control of MBB. Pedio wasps are introduced as soon as MBB larvae are spotted. If properly released, they control MBB the entire season.

The Pedio wasp makes its entire living on MBB and can detect it from at least a mile, based on experiences with my own releases. Unfortunately it doesn't usually overwinter, though it did overwinter one year in Virginia. The locations of MBB are also flagged so that they can be released near the hot spots. The MBB can reproduce faster than the Pedio wasp when temperatures are cool, but by mid June and later, the temperatures are high enough for the wasp to significantly overtake the MBB.

When the wasp (about the size of a pin head) parasitizes the MBB it lays its eggs in the larvae. Shortly thereafter the larvae become immobile and turn brown. Depending on the source, the Pedio wasp costs about $20 per 1,000 wasps plus FedEx delivery. Some insectaries sell the mummies, the brown parasitised larvae of MBB.

These are placed around the plants for later emergence of the adult wasps. Adult wasps should be released either early in the morning, or at about dusk. In either instance, the foliage should be dry. The key to the success of this approach is early detection of MBB and timely release of the Pedio wasp.

Bean Weevil (*Acanthoscelides obtectus*)

Bean weevils lay their eggs on bean seed, producing larvae which develop and emerge through a small hole in the seed. Bean weevils can quickly destroy a batch of stored seed if the seed has not been inspected for a while. Beans can become infested in the field so it is important to harvest and process the seed promptly. Threshed seed should be bagged and stored promptly.

DISEASES

There are a large number of diseases that affect beans, though their prevalence is related to location (certain areas of the country), climate, weather, and types of insect vectors. Many of the disease problems can be controlled or reduced by practicing a three-year crop rotation. Only the diseases which are more serious in the Mid-Atlantic and southeastern U.S. are covered here.

Anthracnose (caused by the seed-borne fungus Colletotrichum lindemuthianum): This disease may be common and widespread in the eastern and southern U.S., especially during periods of cool, humid, rainy weather which favors disease development. The plant can be attacked at any stage of growth and the disease spores are carried by wind, soil-splash from rain, gardening tools, and farm implements. The disease is characterised by black sunken lesions up to ½" in diameter (with diffuse margins) on the stems, pods, and cotyledons. Leaf veins on the under side of the leaf may turn black. During wet weather the spots may be covered with a salmon-coloured ooze. The fungus overwinters in bean seed and residues of diseased plants. For this reason, proper cleanup of the gardens or field, and prompt composting of crops residues is important after harvest.

Rust

Bean rust is most common on mature plants and is characterised by white spots or flecks on the underside of the leaf which turn into rust-coloured lesions after several days, followed by yellowing and drying of the leaf which later falls off. The symptoms mostly affect the leaves. The fungus overwinters on diseased plants left in the field. The disease can be controlled with sulfur dust but the disease can also be minimised by use of crop rotation and avoidance of overhead watering in late afternoon or evening. (Sulfur dust is an allowed material for certified organic production; however, check with your certifier for specific regulations.)

Powdery Mildew

Powdery mildew usually appears late in the season and is common in the southern states and along the coast. Symptoms first appear as small, grayish-white powdery spots which later coalesce to cover the entire leaf. Young leaves may curl, yellow, and drop if the disease is severe. The disease spreads most rapidly when humidity is high. Though the disease can be controlled with sulfur dust, a small amount of disease on the leaves is not harmful. Avoid irrigating plants from above on overcast days or late in the day so that moisture has an opportunity to evaporate from the leaves.

Bacterial Blights

There are four different bacterial blights that affect beans. They all have somewhat similar symptoms and all are favored by wet weather. Most blights start as water-soaked areas on the leaves and stems. The spots later turn brown and coalesce and finally kill the leaf and injure the pod. The bacteria are spread by rain, wind, infected crop residues, and diseased seed. The four blights (listed in order of importance) are: (1) common blight (Xanthomonas phaseoli); (2) halo blight (Pseudomonas phaseolicola); (3) fuscous blight (Xanthomonas fuscans); and (4) brown-spot blight (Corynebacterium flaccumfaciens).

Of the four blights, common blight is the most serious, and is favored by warmer temperatures: it is especially a problem in the Mid-Atlantic and South. Some bean varieties have tolerance to common blight, but none have a high level of resistance to this disease. In contrast, a number of bean varieties have resistance to halo blight and their use is recommended where possible. It should be cautioned though, that disease studies can be difficult to interpret: investigators often report differences in their own tests from year-to-year.

The most effective means of control is using seed from blight-free areas where fields are surface irrigated and rains are infrequent. Few areas of the country meet this requirement, for example areas of California and the Snake River Valley of Idaho. Another effective means of control is use of resistant varieties, but resistance to disease is often a matter of degree rather than an all-or-none phenomenon. For regional seed production based in the Mid-Atlantic and Southeast, it may be difficult or impractical to implement the first two control measures. Therefore, for our region, the first line of defense includes recognition of diseases symptoms, use of a three-year crop rotation, and timely removal of crop residues. If the residues are infected they should be buried.

Crop rotation in itself may or may not provide the desired results, but is strongly recommended as a common cultural method. Additional control measures include the following: (1) avoid overhead irrigation if possible; (2) maintain air circulation around plants through use of weed control and wider plant spacing; and (3), don't work among wet plants. As a last resort, application of copper and sulfur-containing sprays can be effective in controlling bacterial blights though there may be a reduction in crop yield.

Mosaic

All three mosaic viruses are spread by aphids. Symptoms of mosaic are typified by some form of yellow and green mottling of the leaf between the veins which remain green. The infected leaves may be puckered, deformed, and rough in texture, and the shoot tips may die back. The disease is spread by contact of the leaves with each other, farm implements, and handling healthy plants after working with diseased plants. Aphids, which spread the disease, seem to be more of a problem where the aphids have become resistant to crops sprayed with agrochemicals. In an organic system where ladybird beetles flourish, aphid populations may not reach such high levels.

REFERENCES

David Burton, "Undercooked Beans Can Cause Illness", University of Missouri, Retrieved on 2007.

10

Tomato Seed Production

The tomato (*Solanum lycopersicum*) is a plant in the Solanaceae or nightshade family, as are its close cousins tobacco, chili peppers, potato, and eggplant. The tomato is native to Central, South, and southern North America from Mexico to Argentina.

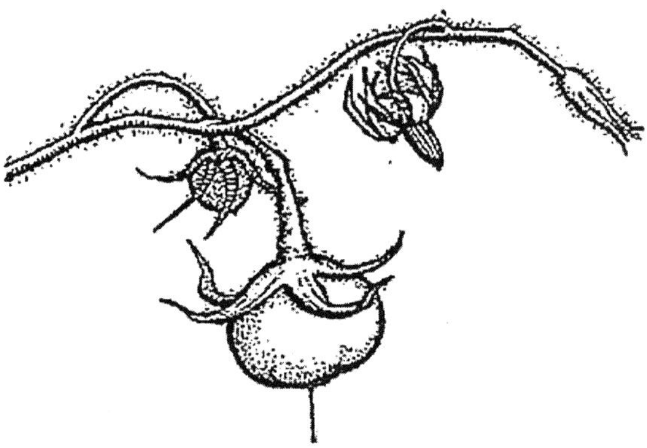

It is a perennial, often grown outdoors in temperate climates as an annual, typically reaching to 1–3 m (3 to 10 ft) in height, with a weak, woody stem that often vines over other plants.

BOTANICAL CHARACTERISTICS

Botanical classification of cultivated species of tomatoes:

— Family: Solanaceae (nightshade family)

— Common tomato: Lycopersicon lycopersicum (formerly Lycopersicon esculentum)
— Currant tomato: Lycopersicon pimpinellifolium

The common tomato produces fruits in a large variety of shapes, colors, and sizes whereas the currant tomato, which has not been domesticated to the same degree, produces red or yellow fruit approximately ¼" to 3/8" in diameter. The common tomato will cross readily with the currant tomato even though the two are different species.

Roots

Left to grow naturally, tomatoes have a strong taproot which can grow as long as six feet, but because tomatoes are normally transplanted, the taproot is broken and the plant develops a fibrous rooting system. Most of the lateral roots develop 2 to 10" beneath the soil surface. The laterals can grow out a distance of two feet before growing downward to a depth of four feet.

Flowers

Tomato flowers are borne in clusters that resemble a raceme. Technically they are classified as a monochasial cyme. Indeterminate varieties of tomatoes have a flower cluster that occurs at every third internode along the main axis. Indeterminates grow indefinitely provided that environmental conditions remain favorable. Determinates have flower clusters that occur at every other node, or as frequently as every node. This growth arrangement continues along the axis until a terminal cluster is formed and growth of that particular branch ceases.

Determinate plants are short-vined, and are sometimes termed "self-pruning." Because they have a larger number of flower clusters per length of vine, the plant tends to fruit and mature a crop within a briefer period of time. In most varieties of tomatoes, the number of flowers in a cluster is typically four to five. In the small-fruited cherry and currant tomatoes, the number of flower per cluster can range up to eight to twelve. One of the more interesting varieties is 'Riesentraube' which can have dozens of flowers per cluster, though not all of these bear fruit.

Pollination and Fertilisation

Pollination is the process of transferring the pollen from the male part of the flower (anthers) to the female part, called the stigma (pollen-receptive surface of the pistil). Fertilisation is the process of union of the male gamete in the pollen with the egg to form the zygote which becomes part of the seed. When the stigma becomes receptive to pollen, anthers do not shed their pollen until about 24 to hours later. During this time, cross-pollination can occur, but after the anthers shed their pollen, the flower is more likely to be self-pollinated. Tomatoes are predominately self-fertilised, but not exclusively.

A large amount of cross-fertilisation can occur under the right conditions depending on the environment, microclimate, and biodiversity in the growing area. Pollen germinates (develops a pollen tube) when the temperature ranges from 50 to 100°F (10°C 38°C). The optimum temperature is 85°F (29°C). Pollen can be killed at 95°F (35°C) or above, depending on the variety. Pollen tube development may also be impaired by high temperatures. Even the optimum temperature of 85°F (29°C) fertilisation has been reported to take as long as 50 hours.

Some of the larger-fruited varieties of tomato do not set seed well during periods of extended hot weather, in the mid-90's or above, and especially if the night-time temperatures don't drop below 70°F 21°C). Some of the more well-known large-fruited heirloom tomatoes have difficulty setting fruit in the hot arid Southwest or deep southern states. On the other hand, small-fruited varieties have a significantly better tolerance to both high and low temperatures, and can be more successfully grown temperature extremes.

Determinate and Indeterminate Varieties

Determinate varieties are short-vined plants on which blossoms and fruit develop about the same time. Indeterminate varieties are long-vined plants which bear fruit continuously. They should be caged or staked as noted below. Some varieties are semi-determinate.

Seed

Tomato seed develops in a mucilaginous gel which has germination inhibitors. During the process of seed extraction and fermentation this gel is broken down. After the seeds are washed and dried, the seeds are normally tan or light brown in color with a pubescent covering (fuzz). Tomatoes are unique among the Solanaceae in that they are the only seed that is pubescent. The number of seeds per fruit typically ranges from about 150 to 300 or more seeds per fruit. Tomato seed germinates in the range of 50 to 95°F (10°C to 35°C). The optimum range is 60 to 85°F (16°C to 29°C), and optimum germination occurs at 85°F (29°C). The Federal standard for germination is 75%.

Isolation Distances

The subject of isolation distances in tomatoes has either been ignored in most seed production guides, or the information is incorrect. The issue is also controversial. There are a variety of reasons for the controversy, mostly stemming from a lack of understanding of what factors are important for determining isolation distances.

The most important point for the seed grower to remember is that isolation distances should be understood within the context of the environment in which the crops are grown. The manual on isolation distances helps the grower understand that context. In addition, the basic principles of pollination ecology are explained, along with practices that can be used to modify isolation distances according to your growing conditions. It also contains a chart of recommended minimum isolation distances, and factors to consider when making modifications to the recommendations.

One key to understanding isolation distances in tomatoes is knowing that the tomato originated in South America, largely in the area of Ecuador and Peru where it was (and is) a plant pollinated by wild solitary bees. During the domestication of the wild tomato, it gradually moved out of its original geographic range, and out of the range of native pollinators. For this reason, the tomato came under increasing selection pressure to become self-pollinated. So, although tomatoes have been domesticated to a great degree, and are now mostly self-pollinated, they retain some of their ancestral capacity for cross-pollination, depending on the species and variety.

Traditionally, most seed saving guides indicate that tomatoes are self-pollinated. This is essentially true for modern varieties, but under certain conditions for certain varieties, considerable cross-pollination of tomatoes can take place. Even a small amount of cross-pollination over a number of years can lead to the loss of one or more characteristics unique to a variety. Therefore it is essential that varieties be isolated from each other in order to obtain pure seed. There are many factors that affect the amount of cross-pollination. These include:

1) variety characteristics such as flower structure;
2) environmental variables such as wind movement, light intensity, day length, and carbon-nitrogen ratio;
3) types of pollinating bees present and their behaviour on the blossoms;
4) isolation distance;

5) presence of barrier plants;
6) planting patterns such as row or block planting;
7) number of varieties;
8) number of plants of each variety; and
9) regional or bioclimatic factors.

Day length can effect the length of the style, as can heat and low humidity. Long day length, periods of high temperature, and low humidity cause the style to elongate which in turn favours cross-pollination.

Generally speaking, most modern varieties can be isolated from each other by a relatively short distance. This is because most modern tomato varieties have a blossom structure in which the length of style does not exceed the length of the anther cone. This arrangement of flower structure favours self-fertilisation. Older tomato varieties, potato-leaf varieties, and large-fruited, beefsteak-type tomatoes (including varieties with double blossoms) tend to have an "exserted style," meaning that the style protrudes beyond the anther cone, typically by a millimeter or more.

This arrangement of reproductive parts favours cross-fertilisation by pollinating wild bees. Some garden tomatoes have retained some of the characteristics of their wild ancestors. These characteristics may be found in many cherry tomatoes and currant tomatoes. These types have a blossom structure such that the style protrudes considerably beyond the anther cone If you are not familiar with the varieties you intend to grow for seed, you should be conservative in establishing isolation distances until you know the variety characteristics.

Also when growing different varieties for seed, it is good to get in the habit of examining the flower structure of different varieties so that you can begin to get an idea of the range of variation and the potential for outcrossing. These distances can be modified on the basis of variables such as the presence of other pollen sources, or the presence of physical barriers.

Less separation is required if seed is collected only from plants in the center of block plantings. A tall barrier crop such as ornamental sunflowers is recommended. Pollen-producing crops such as squash and cucumbers are useful in providing bees with an alternate source of pollen, though this pollen is available mostly up until mid-day. Borders of perennial flowers are very useful for attracting bees, especially those flowers that have open, exposed nectaries, for example, members of the Composite Family.

In making decisions about isolation it is important to take into consideration plantings made in neighboring gardens or farms. If a nearby garden or market garden is a problem for maintaining isolation, consider donating the same seed variety to your neighbour.

POPULATION GENETICS

Plants grown for seed can become subject to "inbreeding depression". This is a condition where the genetic diversity of a variety becomes so narrow, or bottle-necked that the variety loses some of its vigor, fitness, and adaptability as a result of too few plants being saved for seed over one or more generations. Generally speaking, inbreeding depression is generally not considered an issue in "selfers" (plants that are self-pollinated). Because tomatoes are predominately self-pollinated they can be treated as "selfers", but even selfers can be subject to inbreeding depression if too few plants are saved for seed.

Because population geneticists have varied responses to the question of the minimum population sizes needed to prevent inbreeding depression, this topic needs more study. Like the topic of isolation distances, the devil is in the details. The minimum population size depends on the variables involved. That said, there are some general guidelines. For plants such as tomatoes, no fewer than 20 plants should be grown for seed, though for a highly uniform (homozygous variety) as few as 12 plants might be required (but that would be an unusual situation). According to plant breeder John Navazio, as many as 80 to 100 plants should be the minimum population size for most selfers.

If you are growing a fairly uniform variety of an established commercial tomato that is available from a number of sources, 20 plants can be considered the minimum. If you are growing an uncommon or rare heirloom variety that has more inherent genetic diversity, you should be growing at least 40 plants, and perhaps as many as 80, if space and resources allow. The question to consider is: "Who is maintaining the diversity of the stock seed?" If what you are growing represents a small part of a larger genetic pool, the minimum population size can be smaller. The seed company that provides the stock seed may be maintaining seed based on a larger population size or pooled lots of seed or larger grow outs.

CULTURAL CHARACTERISTICS

Sowing Seed and Handling of Transplants

Sow seeds six weeks (for determinate varieties) to eight weeks (for indeterminate varieties) before the last frost date in your area. The best planting medium is a sterile seed-starting mix. Good drainage and aeration are important for starting seeds. Plant seed ¼" deep in shallow flats or cell packs and maintain soil temperature in the range of 75-85°F (24-29°C). Under optimum conditions, most varieties germinate in five to fourteen days, but under less than ideal conditions, germination may take at least three weeks.

When the seedlings have produced four leaves (two seed leaves and two true leaves), transplant to 3" pots or large cell packs. Seedlings in plastic pots require much less water than those in peat pots. After transplanting, keep seedlings at a lower temperature at night, 50-55°F (10-13°C), to promote earlier flowering in some varieties. Day temperatures should rise to 75-85°F (24-29°C) to promote rapid growth. Expose plants to light and air currents as much as possible to harden the plants and promote stockiness.

If fluorescent lights are used, keep the seedlings no further than two to four inches below the lamps, otherwise seedlings may become leggy. To keep seedlings stocky indoors, direct a gentle breeze from a low-speed fan toward the seedlings for several hours a day. Water sparingly, but do not allow the growth to become checked. Fertilise with complete, soluble organic fertiliser or fish emulsion if leaves become yellow and/or purple. Leaves with a purple coloration may indicate phosphorous deficiency. Yellowed leaves may indicate nitrogen deficiency. For transplanting to the garden, average soil temperature at 8 a.m. (4 inches deep) should be above 60°F (16°C). To facilitate drought resistance and a larger root mass for nutrient absorption, set the plants deep in the topsoil to encourage development of adventitious roots that develop along the stem.

Culture Methods (support of vines)

In the Mid-Atlantic and South it is not advisable to grow indeterminate (and some determinate) tomatoes on the ground for several reasons. Ground culture promotes disease by placing foliage in close proximity to soil-borne pathogens; interferes with good air circulation around foliage; promotes fruit-rot (anthracnose); and subjects developing fruits to higher temperatures which could potentially affect seed development. Plants should be either staked, caged, or trellised (for example, the Florida weave system). Caging produces the best results, especially for tall indeterminate varieties. Cages should be 18 to 24" in diameter, and at least 48" tall. Cages can be supported by fastening the cage to metal electric fence posts, or 48" lengths of rebar which can be purchased at building supply stores.

Spacing of Plants

Recommended spacing distances depend on the variety and the support system used for the vines. For cage culture, plants should be spaced 24 to 48" apart within the row. Rows should be spaced three to six feet apart. Most indeterminate varieties are spaced 36 to 42" within the row and five feet between rows. Closer spacing generally produces higher total yields than wide spacing, but spacing which is too close may promote disease by interfering with air circulation. Before determining spacing, you should be familiar with the width and height of the varieties you plan to grow. Certainly soil fertility, water, tilth, and variety characteristics all have a relationship to the final plant size.

Factors Affecting Yield

Too much mulch on the soil in the spring may delay growth by preventing the soil temperature from rising enough to support active root growth. Starting at the middle or end of June, apply a deep mulch around plants to help conserve moisture and increase yield. High levels of phosphorus are necessary to produce good yields, but too much nitrogen added after transplanting will delay flowering. Pruning and staking increase earliness to fruiting at the expense of yield. Pruning of determinate varieties should be avoided or kept to a minimum.

Diseases

Early determinate varieties generally have poor disease resistance, however disease is not usually a problem until mid-season. Thus disease resistant varieties should be planted for a sustained harvest. Leaf blight diseases such as early blight and alternaria begin to appear about mid-

July, and plants are more susceptible once fruit production begins. To reduce or prevent disease problems, use resistant or tolerant varieties and rotate tomatoes to different parts of the garden each year, using a minimum three-year rotation.

Tomatoes should not be grown in soil which has previously been used to grow peppers, eggplants, or potatoes during the previous three years. Fusarium wilt, a disease caused by a soil fungus, is common in the Mid-Atlantic region, during mid- to late-season. Where fusarium wilt is present, a minimum six-year rotation is recommended. Do not plant eggplants, peppers, or potatoes in wilt-infested soil during the rotation period. Use resistant varieties if space does not permit a sixyear rotation or if there is a history of fusarium in the soil.

Avoid planting tomatoes near walnut trees to avoid "walnut wilt." Early blight and anthracnose are common in the Mid-Atlantic region, and are favored by hot, humid conditions. Late blight is more common in inland regions at higher elevations, especially during the spring and fall. Blossom-end rot is prevented by ensuring an adequate level of soil calcium and steady moisture. Do not use fungicides. Fungicides may mask important information about variety performance. When fungicides are absent there is a good opportunity to identify and select plants which have greater resistance to disease.

Greenhouse and Solar Greenhouse Notes

Greenhouse-grown tomatoes require pollination for good fruit set: vibrate the blossoms with an electric toothbrush or tap the flower cluster several times with a pencil. Fans can be used in the greenhouse to vibrate the blossoms during the wintertime when greenhouses are not vented. The daytime greenhouse temperature should not exceed 90oF (32°C), and night temperatures should drop below 70°F (21°C), but not lower than 55°F (13°C). The optimum night temperature is 59-68°F (15-20°C). At 40°F (5°C) some tomato varieties show tissue damage not readily visible to the eye. Greenhouse tomato pests can be controlled with insecticidal soap up to one day prior to harvest. Whiteflies, winged-aphids, and leaf miners are attracted to and trapped by sticky yellow traps.

SELECTION

Selecting the Plants

There is usually no need to rogue plants, provided the seed crop is being grown from well-selected stock seed. However, in dealing with ethnic or heirloom varieties it is not unusual to see more variability in a seed crop. Many ethnic groups and gardeners maintaining family heirloom seed do not place the same premium on uniformity valued by the commercial seed trade. Variability in ethnic and heirloom varieties is not necessarily bad. In fact the variability may be an important source of genetic diversity. So, in general, you may expect to see more variability in ethnic and heirloom varieties, but plants that are obviously off-type should be rogued. If you are growing seed of an ethnic or heirloom variety that is new to the seed trade, the seed company should provide you with some idea of what type of variation to expect.

Selection of seed plants begins at the seedling stage and continues before the plants have flowered. Remove any weak plants or those with any off-type foliage before these can pollinate with any other plants. Pay attention to the foliage of the plant and be aware of any increased or decreased susceptibility to disease, insects, or environmental conditions. If you see a plant with some unusual or potentially useful characteristic, you may want to transplant it and isolate it for future evaluation.

Once the plants begin to bear fruit, examine each one for trueness to type. Instead of focusing on each individual fruit, you should look instead at the overall performance of each plant. Does each plant have typical fruit color, shape, size, and internal fruit characteristics? If a significant number of fruits on a plant fail to meet overall performance requirements, the plant should be pulled, otherwise it is too late for roguing plants once the fruit has been picked.

Selecting Fruit on the Plants

In selecting fruit for seed, choose fruit which are true-to-type including typical color, shape, size, and internal fruit characteristics. In other words, do not use off-type, misshapen, or diseased fruit. Fruits which are bruised, or fruits which have small cracks are useable for seed, but plants which have a lot of cracked fruit should be rogued before fruit is harvested, otherwise you may end up selecting for cracking. It is okay to harvest both early and late fruit, but don't attempt to harvest fruit from plants which are obviously in stress or decline. During seed processing, keep an eye out for fruit which has undesirable internal qualities.

HARVESTING

Tomatoes should be harvested at the "dead-ripe" stage, but they should not be left on the vine so long that there is evidence of decay. They should have developed full color, and the fruit wall will have softened enough to have a slight "give." During very hot weather, harvesting tomatoes about two days before they are dead ripe so that they can continue to ripen off the vine in the shade. If your growing season is shortened prematurely by frost you can harvest the fruit and allow it to ripen off the plant. Interestingly, seed harvested from immature fruit at the "breaker stage" (when it shows the first blush of colour) will often germinate well, but such seeds will not retain their viability as well as seed harvested from fully mature fruit. If you do harvest seed from slightly immature fruit, it should be labeled as such.

In the southeastern U.S. daytime temperatures can climb into the high 90's. Once fruits are harvested they should be kept in the shade because the effect of direct sunlight on the fruits for long hours could potentially raise internal fruit temperatures high enough to damage the seed. When tomatoes are grown for seed in the Deep South it may be useful to partly shade the plants from intense afternoon sun. This can be done by growing them near a line of trees that have a large canopy with few branches on the lower trunks.

Fruits ripening on the vine at temperatures above 90°F (32°C) may produce lower germination seed. This problem most likely occurs in varieties that do not have good foliage-cover. Normally, this is not a severe problem and doesn't require corrective action, but may explain lower germination of seed extracted during periods of hot weather.

METHODS OF SEED EXTRACTION

There are basically three methods of seed extraction: (1) juice and seed extraction, (2) acid extraction which is not recommended, and (3) extraction by fermentation, which is the preferred method. Fermentation is the preferred method because it is a natural process that is least harmful to the seed and can destroy bacterial canker and other seed-borne diseases. Fermentation should be a controlled process. Though not difficult to do, it can be done incorrectly, in which case the ferment produces a bad smell and an overgrowth of white fungus which can stain and damage the seed.

FERMENTATIVE EXTRACTION

The preferred method for producing commercial-grade seed. The best quality seed is obtained by fermentative extraction. The process basically consists of breaking or mashing up the fruit into pulp, seeds, and juice, and then pouring the mixture ("mash") into a large container where it ferments for a period lasting usually three days. After fermentation is complete the seed is separated by washing, and then the seed is dried. Though the process is quite simple there are some important details for performing the process properly.

Methods for Mashing the Fruits

The first step in extracting the seed is to reduce the fruit to a coarse mash. Before starting this it is sometimes useful to core the tomatoes first because the cores are heavier, they don't break down well during fermentation, and they don't wash away easily during the washing process. Tomatoes that have dirt or other debris stuck to the skin should be rinsed off before processing. There are all sorts of methods that can be used for mashing the fruits, some more fun than others. Be creative. There are lots of possibilities, and if you develop a new method let me know. Some suggestions include the following:

Bucket and Masher Method

Cut the tomatoes into wedges or sections and add them to 5-gallon plastic buckets, filling the buckets no more than two-thirds full. Lids are recommended to keep out fruit flies which can become a nuisance when fermentation is done indoors. Five-gallon plastic buckets are available at most building supply stores. Larger containers such as feed troughs are available at farm supply stores. You may also want to purchase a hand mortar mixer which looks like an oversized potato masher. This is useful for mashing up the cut-up sections.

Alternatively you can mash up the sections by squeezing them through your hands. If this seems too gross you can use the mortar mixer to mash them. This secondary mashing method isn't always necessary if the tomatoes break apart easily, but the fermenting mash will get a better start and will be more uniform if the fruits are well broken down at the outset. The advantage of this method is that it is probably the easiest method for producing small lots of seed ranging from two ounces to a few pounds. Make sure that you label the bucket with the tomato variety, and remove old labels when the buckets are cleaned.

Wheelbarrow Method

This method was used by Glenn Drowns, one of our former growers, who used it for extracting seed of sweet peppers. Fill a wheelbarrow about half-full and chop the tomatoes with a hoe. Make sure the wheelbarrow is on level ground. The disadvantage of this method is that the tomatoes may not be equally chopped. This method is for softfruited, medium or large tomatoes: you can't chop cherry tomatoes with a hoe. The advantage of this method is that if you are allergic to tomatoes you don't have to handle them directly. Though rare, some people can develop rashes or allergies from frequent contact with tomato fruit and vines. After the tomatoes are chopped, transfer the mash to a fermentation container.

Swimming Pool Method

I have not used this method, but suggest it as a possibility to explore. This method has been used successfully by a grower for a small southwestern seed company. Fill a plastic swimming pool one-third full of tomatoes and step on the tomatoes as though they are grapes to be smashed into wine. At the end of fermentation, water can be trickled into the pool at a slow rate (so as to not agitate or float the seeds on the bottom). The water will clarify, but pulp will need to be scooped or picked out. Alternatively, the mash can be transferred to 5-gallon buckets or other large containers. This extraction method can be a great activity for kids. The disadvantage is that the kids will need to wash their feet and legs thoroughly before moving on to the next variety, and the pool will need to be thoroughly cleaned. Without thorough cleaning, seeds and diseases of different varieties can be mixed.

Hardware-cloth Grater Method

Build a wood frame using two-by-four lumber and mount ¼", 3/8" or ½" hardware cloth on the frame. (The frame can serve double-duty as a soil sieve.) Place a large tub or basin under the frame. If the tomatoes are soft, you can rub or grate tomatoes over the hardware cloth. It may be necessary to cut the tomatoes first, halfway between the blossom and stem end.

Other Methods

The following are other variations to consider, though for small seed lots are easy enough just to cut tomatoes into sections, break them up, and ferment them in quart or half-gallon jars.

Zip Lock Bag Method

This method comes courtesy of grower Merlyn Niedens who uses this method for processing small amounts of seed. Purchase heavy-duty, freezer-quality, quart or gallonsized zip lock bags. Write the name of the variety on the side of the bag. Place the tomatoes inside the bag and crush them. The bag or bags are placed inside a large basin or 5-gallon bucket. These bags are massaged or squeezed at least twice daily to stir the ferment. After squeezing the bag several times, partially open the top of the bags so that gas can escape. Be sure to position or support the bag so the ferment doesn't spill. When fermentation is complete the contents are placed in the bottom of a bucket to be washed. When done, wash out the bag and hand upside down to dry.

Food Processor Method

This doesn't offer an advantage over other methods, except that it can be useful for cherry or currant tomatoes. It does reduce the mash to a finer texture, but the fine texture is not necessary for good fermentation.

Use a food processor (not a blender) with a blunt plastic blade and set on a slow setting, or process with a quick turn on/off motion. This is a very fast and efficient method for processing small amounts of fruit. Be sure your food processor is equipped with a 1/8" dull square-edge plastic blade. Do not use a metal blade which will damage the seed. Cut tomatoes into wedges; drop into the food processor until about half full, and process at a slow speed until the wedges are broken apart.

Squeezo® Strainer Method

Two products are produced by this method: the juice and the "pumice" which consist of "dry" pulp, skin, and seeds. The seed is separated from the pumice by washing the seed with water until the seed is clean. The advantage of this method is that you can save the juice to drink, but the disadvantage is that it is a slower process, and fermentation isn't used (unless you save some of the juice for fermentation). There is a possibility seed may be bruised by this method, so if this method is used, compare the germination of seed extracted this way with seed (of the same lot) processed by the bucket and masher method. Be sure to clean the strainer thoroughly between processing varieties. Seed may become stuck in small crevices.

Corona Grain Mill Method

The Corona grain mill can be used to pulp tomatoes that have been cut into sections. The important point to remember is that the milling plates have to be set far enough apart so that they do not grind or abrade the seed.

A disadvantage of this method is that seed can become stuck in crevices of the mill plates or grinder and need to be cleaned out, a concern if you are processing more than one variety.

Adding Water to the Mash

If bacterial canker is a problem, it is best to not add water to the mash because dilution of the mash reduces the effectiveness of controlling canker. If canker is not a problem, it doesn't hurt to add a small amount of water, usually not more than 10 percent of the volume of the mash. Adding water makes it easier to stir the mash, and allows more circulation of ingredients since the mash is less dense and the rising bubbles have an easier path to the top of the liquid.

Fermentation Process

After the fruits are reduced to a mixture of sections, pulp, juice and seed, the next step is fermenting the mash. This happens naturally without any intervention. The combination of naturally occurring yeast and sugars in the mash is enough to start the process. Fermentation has begun when small gas bubbles are seen rising from the mash, usually within twelve hours. Fermentation is complete when gas production slows or stops, when the mucilaginous material around the seed is digested, and seed has settled to the bottom of the container.

A very important step, often not mentioned in seed-saving literature, is that the mash should be stirred at least twice, and preferably three times a day. This helps the ferment to work evenly, helps release the seed from the pulp and its mucilaginous coat. More importantly, stirring prevents the growth of a white scum on top of the mash. This white scum, often accompanied by a bad smell, indicates the overgrowth of a fungus which can injure or discolor seeds. If you see this overgrowth you need to stir the mash more frequently.

Fermentation Temperature

The speed of the fermentation process is largely dependent on the temperature. If the temperature ranges between 75 to 80°F (24 to 27°C), fermentation should be complete within 48 to 72 hours. To control bacterial canker, the fermentation should last 96 hours (4 days). This requires a temperature of about 65°F (18°C). This long fermentation can cause injury to the seed, such as premature sprouting. Therefore as a compromise, tomatoes should ideally be fermented at close to 70°F (21°C) as possible.

This may be difficult to achieve in practice during the summer unless fermentation occurs in an air-conditioned environment or cool basement. At a minimum keep the fermentation temperature below 80°F (21°C) because a fermentation temperature (temperature of the liquid) in the high 80's can cause a 50% loss of germination in 48 hours. A fermentation temperature in the 90's will cause a loss of germination in the majority of the seed. If fermentation occurs outside, make sure that it is done in the shade.

When fermentation is complete, the mash has lost its frothiness when stirred, the good seeds have sunk to the bottom, and much of the pulp has floated to the top.

Washing

When tomato seed is processed on a large commercial scale, the seed is washed in a shaker washer or flume. In small-scale seed production where seed is fermented in containers such as buckets, the first step in washing is to stir the fermented mash, let it settle, and then scoop off much of the pulp floating on top. Next, water is poured into the mash so that the volume is approximately doubled. If you don't add enough water, the specific gravity of the mash may be too high to allow some of the good seeds to sink.

Adding water lowers the specific gravity. This also clarifies the liquid so that you can see the seed more easily. After water has been added, the mix is allowed to settle again. The good seeds will sink to the bottom. After the mix has settled, the container is tilted and the pulp and other debris are poured off the top. The pouring continues until most of the mash is poured off. Lighter, low-density seeds of poor quality can be poured off with the liquid. This washing process is repeated until the water turns clear. Typically it takes anywhere from three to six washings.

When the water comes clear, the contents are then poured into a large strainer (held over another bucket). Excess water is expressed by pressing on the seeds with a large spoon while the seeds are in the strainer. Next the strainer is flipped upside down over a cloth towel or rag. The seeds fall out into a pile, and the pile is then spread and flattened with a large spoon until the patty is no thicker than ¼ inch. The towel itself is usually spread over a screen so that air can reach the patty from the bottom as well as the top. The purpose of the towel is to wick water away as quickly as possible. The patty of seeds doesn't need to be turned or stirred provided there is a steady stream of air flowing over the seeds.

Drying

Once the seed has been washed and made ready for drying, it should be dried as quickly as possible without heat. A fan is useful for this purpose. If the seed is not dried quickly it can sprout or mold. The seed should be relatively dry to touch within 24 to 36 hours. Drying should take place at a temperature of less than 90°F (32°C). Once the temperature reaches 95oF (35°C) damage to the seed can occur. For this reason, seed should not be dried in the sun when the air temperature is much over 80°F (27°C). Seed is best dried in an air-conditioned room with air circulated by fans.

Curing

Curing is a continuation of the drying process. After the seed has been dried set it aside in a dry environment with good air circulation to cure for two to four weeks. During this time the moisture content of the seed is gradually reduced. Seed should be thoroughly cured before being placed in an airtight container. If curing outdoors, make sure that mice cannot get into the seed.

Milling

It is usually not necessary to mill seed after washing and drying since there will be little foreign material with the seed. It may be helpful to screen the seed for size in order to discard small seed.

Seed Yield

Calculating seed yield is an optional step that can be useful for determining seed yields of different tomato varieties. This is important information if you are a seed grower planning to sell small lots of seed to seed companies. This only needs to be done the first time you grow a variety for seed, provided that your growing conditions remain relatively the same from year to year and place to place. To calculate seed yield, weigh the total lot of tomatoes to be extracted for seed, and then record this information.

After you have extracted, washed, and dried the seed, weigh the dry seed and calculate how many pounds of fruit are required to produce an ounce or pound of seed. You will also want to keep a record of the number of plants required to produce a given weight of seed so that in the future, if you want to produce a pound of seed you will know approximately how many plants you will need.

An acre of tomatoes raised for seed may yield an average of 35 to 160 pounds of seed depending on the variety, the density of the planting, the amount of crop saved for seed, fertilisation efficiency, and growing conditions. One hundred pounds of tomatoes will yield approximately 4 to 10 ounces of seed, but this number is strongly influenced by the fruit size, variety, and growing conditions. Seed size and weight varies depending on variety, and averages about 11,480 seeds per ounce. The Federal germination standard is 75%.

Labeling

It is easy to lose a good batch of seed if you have not labeled your seed at every step from extraction through drying. Never assume that you will remember which variety is which. The label should travel with every step of the process from start to finish.

INSECT PESTS

An important thing to remember about insect pests is that just because an insect is feeding on a plant doesn't necessarily mean that there will be an adverse effect on yield. Some plants can withstand quite a bit of defoliation without serious consequences, so a zero-tolerance policy is not advocated. Instead, a policy of damage assessment based on cost-benefit is more ecologically sustainable. Cost-benefit analysis involves awareness of what role the pest is playing in the ecosystem. Generally speaking, if you have created a growing area with rich biological diversity, your pest problems are likely to be minor.

Tomato Hornworms

Tomato hornworms can be controlled by Bacillus thuringiensis, though in most cases it is not necessary to use this unless the outbreak is severe. Handpicking hornworms early in the season often provides sufficient control.

Incorporate biological diversity into the growing area. This encourages the activities of small parasitic wasps to lay their eggs on hornworms. As the wasp larvae develop inside the hornworm, the hornworm becomes lethargic or immobile, and shortly thereafter, the young wasps emerge from the caterpillar. Hornworms with the white egg cases of parasitic wasps on their backs should not be destroyed because the wasps will hatch out and parasitize other hornworms in the garden. It is worth noting that hornworms hatch out into night-flying sphinx moths which are important pollinators of some flower and herb crops.

Fruitworm

Fruitworm, also known as the corn earworm can be controlled by Bacillus thuringiensis. Tomatoes with thick skins, especially cherry tomatoes are much more resistant to fruitworm than varieties with a thin skin.

Flea Beetles

Flea beetles are rarely a serious problem. A simple control measure is to keep flats of tomato seedlings several feet off the ground because the flea beetles have a more difficult time reaching the plants.

CONTROL OF SEED-BORNE DISEASES

Tomatoes are prone to a number of diseases, many of which affect yields noticeably. There are over 20 bacterial and fungal diseases which can affect tomatoes, plus others caused by insects, or nonparasitic diseases, also called physiological diseases (for example, blossom-end rot). Sometimes more than one disease may be present at a time. As a seed grower, the diseases that are of most concern are those diseases which can be carried in the seed or on the surface of the seed coat. Basic sanitation and prevention is always the first step. To control seed-borne diseases, it is important to practice crop rotation, to clean up and compost all plant residues, to maintain adequate nutrition, and to select and wash fruit before collecting the seed. Seed crops should be planted early in the season, rather than late.

By planting early in the season a seed crop may be produced before the onset of warm humid conditions which favor fungus diseases. Plants which show disease should be rogued as quickly as possible to minimise disease spread. Seed should be collected only from healthy plants. Fruit which shows obvious signs of discoloration, fungus growth, mosaic patterns, or other abnormal conditions should not be used for seed (unless there is a risk of losing the variety). The ground should be mulched under the plants. This prevents rain from splashing disease spores onto the plant.

Seed growers should be aware of several seed-borne diseases which may be carried on tomato seed. The most important diseases for the Mid-Atlantic and Deep South are those listed below:

Bacterial Canker

This bacterial disease is not widespread but can be destructive, and it is more common in southern states. As a seed producer, you should be familiar with the symptoms. Fruits infected with canker will show "bird's eye" spots. These are dark rough spots (usually tan or brown, about 3/16" (3mm) in diameter), surrounded either by a water-soaked area, or white halo. Symptoms on foliage are evident as an upward curling of leaves which turn brown and wither but do not fall off. Cankers may develop at leaf nodes and spread downward. The cankers first appear as light streaks which darken and crack; later small cankers may form along the length of the crack.

The pith of the stem may appear yellow and mealy. Canker disease can be eliminated by extracting the seed with fermentation. To destroy canker by fermentation keep the temperature at or below 70°F (21°C) for a minimum of 96 hours. Canker may also be destroyed by acetic acid treatment. Moist, just-extracted seed is soaked for 24 hours in a 0.8% solution made from pure acetic acid (1 ounce acetic acid per gallon of water). If dried seed is being treated, use ¾ ounce acetic acid per gallon of water. If acetic acid is not available it may be possible to obtain a 0.8% solution of acetic acid by diluting white cider vinegar (5% acid strength) with 5-¼ parts water. After acetic acid treatment rinse with water and dry as soon as possible.

Tomato (tobacco) Mosaic Virus

Tomato mosaic is most often seen on greenhouse-grown tomatoes and sometimes field-grown tomatoes. The most common symptom is a mosaic or mottling of the leaves which shows as irregular patches of yellow or light green mingled with green. Leaves may be distorted, blistered, curled, dwarfed, or elongated like shoe strings, depending on the strain of the virus. Infected fruit may not show any symptoms, or may show some internal browning or discoloration. Tomato mosaic virus may be prevented by keeping smokers out of the garden (cigarettes may carry the virus), by dipping your hands in milk each time you handle another plant (milk inactivates the virus), and by using diseasefree seed.

Tomato mosaic virus doesn't get into the seed embryo but is carried on the seed coat and is difficult to remove or inactivate. Seed harvested from infected plants may produce up to 3% infected seedlings. The virus can be attenuated or inactivated by storing dry seed at room temperature for several months to a year. Seed which has been stored several months will usually produce normal seedlings, though the virus may still be present in the seed coat. In addition to dry storage, the virus may also be inactivated by treating seed with 10% trisodium phosphate for 10 minutes followed by 3 water rinses lasting 5 minutes each. Planting the seed directly without later transplanting will also reduce the incidence of the disease.

Fusarium Wilt

This fungal disease is fairly widespread in occurrence, and more prevalent in southern and warm coastal states. Wilting is the most obvious symptom. Typically one stem may wilt, yellow, and die back at the top, presenting a flag-like appearance, or yellowing and wilting may occur only on one side of a leaf midrib. The infected stem may die back or wilt severely before other stems show infection. If you suspect fusarium wilt, it can be confirmed by cutting the stem in cross section.

The vascular tissue of the stem will be discolored, typically showing as a brown ring. Plants showing fusarium wilt should be pulled and the vines should not be composted. Instead they should be discarded in the trash or a location far away from the growing area. The disease is fairly persistent in the soil, and plants in the tomato family (Solanaceae) should not be grown in fusarium-infested soil for at least six to eight years. Control measures include paying strict attention to crop rotation and growing wiltresistant varieties. Hot water seed treatment can be an effective treatment in some instances.

Anthracnose

This fungal disease is a fruit rot found on uninjured ripe tomato fruits, and is common in the Mid-Atlantic and southern states. Another name for this disease is called "ripe-rot". In the early stages of infection, lesions on the fruit appear as small, sunken, circular spots that appear as water-soaked indentations in the skin. These gradually grow larger, up to about ½" in diameter, and develop a dark center because of the black fungus under the skin.

The disease becomes more prevalent, especially in August during hot and rainy weather. The disease organism is prevalent in the Mid-Atlantic and grows best at 80oF (27°C), but it doesn't grow above 95°F (35°C) and for that reason isn't as much a problem in the Deep South. Tomato varieties that are subject to defoliation by other diseases are also more susceptible to anthracnose. Lesions also appear on the stem and foliage, but are easily overlooked. These stem and leaf lesions may serve as infection sources for the fruit. This is why it is important to remove plant debris at the end of the season. A crop rotation of at least four years is recommended.

Early Blight

Early blight is fairly common in the Mid-Atlantic, especially during periods of frequent rain. Infection can occur in the seedling stage, but is most commonly observed on the older leaves of established plants, frequently after fruits begin to set. Symptoms develop as dark spots with concentric rings. The rings spread to surrounding tissue and as the infected leaves die, they become dry and papery.

In addition to causing a decrease in yield, the disease can be carried on the seed. The disease is promoted by heavy dews and frequent rainfall. Maintaining good air circulation around the plants can be helpful, as well as mulching the soil to prevent soil splashing of spores onto the leaves. A minimum two-year rotation and removal of crop residues is helpful in controlling the disease. The disease may be seed-borne, but proper fermentation of harvested seed helps to control the disease.

SEED DISINFECTION

In most cases no seed disinfection treatment is recommended for disease control. Proper fermentation of extracted seed will destroy canker and some other fungus diseases. Long-term dry storage will attenuate tomato mosaic virus. If you find it necessary to harvest fruit from plants affected by diseases such as wilts, blights, etc., and are concerned about the possible transmission of disease you can use the following general seed treatment. Prepare a 10% solution of commercial bleach and soak thoroughly dried seed for 10 minutes with frequent stirring.

Pour off the bleach solution and follow with three, 5-minute water rinses. Dry the seed as soon as possible. This treatment may have a small adverse effect on germination. Note: any seed which has been treated with a chemical, no matter how mild, should be labeled to that effect. In fact, Federal seed law requires that if seeds are treated with any substance they must be labeled as such. Also, any seed which you suspect might be a carrier of disease should be labeled. This is especially important if you exchange seed with others or grow seed commercially.

Seed Disinfection and Organic Certification

Seed can be disinfected by hot water treatment. The process involves immersing the seed in water hot enough to kill the pathogen, but not so hot that it damages the seed. This process is discussed in a separate manual dealing with control of seed-borne diseases. Check with your certifier to assess the different seed disinfection options available.

SEED STORAGE

Long-term Storage

It is very important that seed be thoroughly dry. Freshly harvested seed should be cured for a minimum of three to four weeks before placing in long-term storage. At least one week before the seeds are placed in airtight containers, they should be transferred to an air-conditioned or heated room or other low humidity environment before being placed in a jar with a rubber-gasketed lid. Thereafter the seed may be transferred to the refrigerator or other cool environment. Before opening the jar, always allow the jar to warm up to room temperature. Try to do this on a dry day so that humid air is not introduced to the jar. Never store seed in a closed container unless the seed has been dried thoroughly first.

Short-term Storage

For short-term seed storage, it is not necessary to store seed in airtight jars. Seed can be stored in large envelopes, paper bags, cloth bags, or other non-airtight containers. Porous containers are not recommended for storing seed for long term unless the surrounding air remains cool and dry. Zip lock bags and most plastics allow water vapor to pass through and therefore these materials are for short-term seed storage only.

REFERENCES

Peralta, I. E. & Spooner, D. M., Granule-bound starch synthase (Gbssi) gene phylogeny of wild tomatoes (*Solanum* L. section *Lycopersicon* Mill. Wettst. Subsection *Lycopersicon*), *American Journal of Botany*, 1888–1902.

Smith, A. F., *The Tomato in America*, University of Illinois Press, 1994.

11

Cucumber Seed Production

The cucumber (*Cucumis sativus*) is a widely cultivated plant in the gourd family Cucurbitaceae, which includes squash, and in the same genus as the muskmelon. Though it technically is a fruit, cucumbers are incorrectly widely considered vegetables.

Cucumbers are considered a quick crop because it only takes about 2 weeks to get a seedling ready for transplant. Once the cucumber plant is transplanted to its final spacing, production should begin in about 30 days. Fruit harvest can continue for 90 days or more.

Cucumber plants require 8 - 10 square feet of greenhouse space compared to tomato spacing of 4 - 5 square feet of greenhouse space per plant. This means that only half as many cucumber plants are required to properly fill the same area. Cucumber plants like higher nitrogen levels and can be grown in USED media that has been previously used for growing media for other crops.

The plants like warm, humid conditions. Being a member of the melon family, they grow very well in the weather conditions that are characteristic of the southern part of the U.S. The plants are heavy producers - yields of 30+ lbs per plant from a 4 month crop are possible during the longer days of late Spring and Summer. The plants are all female and the flowers do not require pollination.

SELECTING SEEDS

European cucumber seeds are available for full sized fruit that is usually harvested when the fruit weighs about one pound. "Mini" varieties are harvested when the fruit is about the size of a large dill pickle. These varieties of cucumber seed are specifically produced for greenhouse production. These varieties are capable of producing fruit without pollination. The seeds grown within the fruit are sterile, and immature. They will not produce another generation of plants.

Many of the newer varieties of seeds have been bred with resistance and/or tolerance to various diseases that have been a problem to growers in the past. The time of year you choose to grow the crop may determine the size of fruit that you desire. During the long days of Summer, all fruit tends to reach its maximum length. During the short days of Winter, the fruit tends to mature at a shorter length. Some growers will select a shorter variety during the Summer months and a longer variety during the Winter months. Their customers receive fruit that is very similar in length throughout the year. Some markets request different sizes of fruit and you may need to make your seed selection based upon the fruit length requirements of your market.

PLANTING THE SEED

Seeds are usually started in 2" or 3 1/2" Net pots and are ready for transplant in 10 days to 2 weeks. When planting the seed, it is best to lay the seed flat. The taproot will extend out from one end of the seed and the growing point of the plant will come from the other end. If you plant the seed with one end pointed down, there is a good possibility that you will have planted about half of your seeds upside down. Lightly cover the seed with about 1/4" of media. Planting the seeds too deep can delay germination and increase the possibility of losses caused by "damping off" fungus.

Once you have finished planting the seeds, thoroughly wet the media with plain, room temperature water. Cover the seed flats with a single sheet of newspaper. Mist the newspaper with plain water until the paper is completely wet. The newspaper will become a "blanket" over the seed flats. It will help to stabilise the temperature around the seeds. It will allow a slight amount of air circulation. As the paper dries out, it will serve as an indicator of when to water the seed flats again.

If you are starting your seedlings in a greenhouse, the newspaper will also serve as a shade clothe to keep the temperature of the growing media within the proper range. Temperatures about 85o F. can "cook" the seeds. When sunlight shines on any dark growing media, it absorbs heat and the temperature of the media can rise much higher than the air temperature. The best germination temperature is 75can "cook" the seeds. When sunlight shines on any dark growing media, it absorbs heat and the temperature of the media can rise much higher than the air temperature. The best germination temperature is 75 can "cook" the seeds. When sunlight shines on any dark growing media, it absorbs heat and the temperature of the media can rise much higher than the air temperature. The best germination temperature is 75.

Although these seeds are expensive, you can expect nearly 100% germination in 2 - 4 days. Water with plain water until the seeds have germinated. The newspaper should be removed once the seeds have germinated. Once the seedlings have emerged from the media, apply Mycostop and Guardian nematodes. Use 1/2 strength fertiliser solution adjusted to a pH of 6.5 - 6.8 for the remainder of the seedling stage.

Transplanting

Cucumbers need 8 - 9 square ft. of greenhouse space per plant. This is double the spacing required for tomato plants. As an example, if your greenhouse was 30′ x 120′ (or 3600 sq. ft) it would hold 400 - 450 cucumber plants. In this example, the plants would be set out in 5 rows of 80 - 90 plants per row.

Two support wires are positioned over each row, about 28" - 30" apart. Assuming the rows are running North to South, we will call the support wires over each row, the East wire and the West wire. Unlike tomato plants, cucumber plants are brittle and will not be "leaned and lowered" when they reach the support wires. As the seedlings begin to grow, a plant clip is used to attach a support string to each plant. Every other plant will be attached to the East support wire, with the alternate plant being attached to the West support wire

The support strings are attached to the support wires on a slight diagonal. As an example, all strings attached to the East wire would be attached to the support wire 18" - 24" North of the vertical position of each plant. All strings attached to the West wire would be positioned 18"- 24" South of the vertical position of the plants. This diagonal angle will help the developing fruit hang away from the main stem of the plant.

The initial growing temperature for cucumbers is 68°F. nights and 80°- 85°F. days. They will tolerate higher temperatures if the relative humidity is 70% or more. The night temperature can be lowered to 65°F. when the plants reach the support wire.

Initial Pruning and Training

As the plants begin to grow, it is important to allow them

to develop a strong root system. A strong root system will enable the plant to support more fruit and grow quicker. During this phase of their growth, all laterals (suckers), fruit and tendrils should be removed. Continue this pruning until the plant has developed 10 true leaves. The plant will be about 3' high before you allow any fruit to develop. This will take about 20 days from the time of transplanting. Keeping the fruit removed during this stage of growth allows the plant to develop a strong root system. The last few leaves in this stage of growth should develop to nearly 12" across.

As the plants grow, the support string should be twisted around the stem, approximately one revolution every 6" - 9". Always twist in the same direction. The string should be carefully positioned to not damage any developing fruit. If you start twisting in a "clock-wise" direction, continue twisting in a "clock-wise" direction all the way to the support wire. You don't want to change the twisting direction to "counter clock-wise" half way up the plant. This could cause a serious problem when you begin removing the lower leaves from the stem. With proper twisting, you will probably not need to use any more plant clips until the plant gets close to the support wire.

Once the cucumber seedlings have been transplanted, you can begin feeding them a full strength, high nitrogen, fertiliser solution. Maintain the night temperatures at 65°-68°F. at the flower level. The relative humidity should be kept at 70% - 85%. A misting system is beneficial for keeping the relative humidity in this range and as an added benefit; a misting system will help to keep spidermites under control. As the plants begin to grow, you will notice that they require 25% - 50% more water than tomato plants. Generating additional carbon dioxide within the building is also helpful during the daylight hours - when the exhaust fans are not running

Hang yellow sticky ribbons and begin a monitoring program for insect pests. The primary insect pests of cucumbers are whiteflies, spidermites and thrips. These can be controlled using biologicals if releases are made before the pests get out of control.

After 10 Leaves

After 10 true leaves have developed on the plants, you can begin developing fruit. If all conditions have been maintained properly, this should be about 20 days after transplanting. The plants will be about 3′ high at this time. If the plants are setting multiple flowers at a leaf node, you should prune down to only one flower at each leaf. Behind each flower will be a small embryo fruit. Select the straightest embryo fruit to mature. If a fruit is curled or crooked, it should be removed. These curled fruit will never straighten out and should not be allowed to waste the energy of the plant. If you allow multiple fruit to develop on the main stem leaves, later developing flowers will probably abort and you may find that you weekly production swings from peaks to valleys.

No pollination is required for these fruit to develop. The plant is all female and will produce fruit that contains sterile seeds. These seeds will not mature. It is believed that mature cucumber seeds cause some people to "burp". Because of the immature seeds, these fruit are considered "burpless".

As the plant continues to grow up the support string, remove the lateral growth (suckers) and the tendrils. If the tendrils are not removed, you will notice that they can grab hold of a developing fruit and damage the appearance of the fruit. The flowers should be thinned to only one straight fruit for each leaf. Continue twisting the growing point around the support string. At this stage of growth, the plants will be growing nearly 6" per day. The fruit will mature to a harvesting size in about 10 days. During this stage of growth, bottom leaves should only be removed if they are interfering with your watering system or if they are beginning to discolor.

It is important that you maintain adequate water and fertiliser levels for the plants as they begin to set fruit. You should always watch the growing point of the plants and make sure that it remains strong. The stem of the plant should be about 1/2" in diameter. The flowers should be a good color, a good size, and it is good to have at least 2 flowers trying to develop at each leaf node. However, if you are developing more than 2 flowers per leaf node, you may be feeding them slightly more fertiliser than they need.

If you have a conductivity meter, you should compare the conductivity of the incoming solution to the plants with the solution that is leaching out of the bottom of the growing media on a weekly basis. Approximately 10% - 20% of the solution going in the top of the media should be running out the bottom each day. The conductivity of your "leachate" solution should be 10% - 20% higher than what is going in the top.

If the conductivity of the leach solution is:

— Less than the conductivity going in the top - you should increase your fertiliser level.

— Equal to the conductivity going in the top - you should reduce the amount you are watering.

— 10% - 20% higher than what is going in the top - your feed & water is OK.

— Greater than 20% higher than what is going in the top - you should increase the number of waterings per day (and possibly leach the media with plain water to reduce the fertiliser levels in the media).

Plant Approaches the Support Wire

As the growing point of the plant approaches the support wire, you should begin removing bottom leaves - 2 or 3 leaves at a time - at a weekly rate that is equivalent to the number of new leaves being set at the top of the plant. At this point, you have reached a good balance of leaves for photosynthesis and proper production. However, pruning is necessary to maintain proper ventilation within the plant canopy for fungus and insect control.

Excessive pruning at any given time can cause too much moisture loss in the plant and may result in aborted flowers. It is better to remove 2 or 3 leaves - wait 2 or 3 days - then prune again. A bottom leaf that is supporting a fruit should not be removed until the fruit has been harvested. The leaves on the main stem will gradually be removed as the fruit is harvested.

It is important to maintain sufficient foliage on the plant for proper photosynthesis. Excessive pruning of leaves can cause the plant to quit producing flowers. However, it is also important to maintain adequate air movement through the plant to reduce the risk of fungus and insect problems.

Watch the growing point of the plant. It may be necessary to increase water and fertiliser by 20% - 25% during times of heavy fruit load. The growing point should be producing a stem diameter of 1/2" and 1 - 2 flowers at each leaf node. If the growing point of the plant is weak as it approaches the support wire, you may choose to remove flowers from the last 2 - 3 leaves before the wire. This will allow the growing point to regain strength and improve the stem diameter for growth along the support wire.

Lateral Growing Techniques

About 30 days after transplant, the plant should be approaching the support wire. Unlike tomato plants, cucumber plants are very brittle and you will break the stem if you attempt to "lean & lower" the plants. There are several methods of training the plants once they reach the support wires.

The "Trellis" method would have you run additional support wires over the top of the aisle way. The plants are then allowed to grow across the top the aisle way towards to plants on the other side of the aisle. This forms an arch with the fruit hanging down above the aisle.

The "Umbrella" method would have you terminate the growing point of the plant about 2 leaves above the support wire. Two laterals are allowed to grow from the plant (one in each direction along the support wire). These laterals grow along the support wire for about 8", then are allowed to grow back down 3' - 4'. This forms an umbrella shape.

The "Number 7" method will perhaps keep your plants the most organised. Each plant is trained in the shape of the number 7. The growing point of each plant is trained to grow along the support wire in the same direction. The growing point is allowed to grow along the support wire until it is 8" - 12" from the main stem of the plant next to it. As the stem grows along the support wire, 2 laterals (suckers) are allowed to begin growing. The main growing point and the 2 laterals are then allowed to grow downward for about 3'. The plant has now been trained in the shape of the number "7" with 2 additional laterals hanging from the support wire.

Once the original growing point or the laterals have grown down about 3' from the support wire, their growing points are pinched out. Removing the growing points will stimulate the plant to produce more laterals. As the fruit matures on each of the hanging vines, it is harvested. Once all the fruit has been harvested from one of the 3 hanging vines, a new developing lateral is located near the support wire. The old hanging vine is then removed just below the new developing lateral. This opens a space for the new lateral to grow. It is allowed to grow down about 3' before its growing point is pinched out. The cycle is repeated on each of the hanging vines until the crop is terminated.

The "Number 7" method keeps the plants growing in an organised fashion throughout their life cycle. It utilises the vertical space within the greenhouse without blocking excessive amounts of light to the developing foliage. It provides a way to renew the foliage growth, while keeping proper ventilation within the plant canopy. Since each of the 3 growing points are developing at different times, fruit production is maintained in a predictable fashion.

FUGUS AND INSECTS PROBLEMS

Like most plants, cucumbers can also be attacked by a variety of root diseases; such as fusarium, damping off, rhyzoctonia and pythium. The regular use of Mycostop, which is a beneficial streptomyces bacteria that will grow around the roots of plants and help shield the roots from attack from theses diseases, may help to reduce the problems from these problems

Insect pests of cucumbers include; whiteflies, spidermites, thrips, aphids and fungus gnats. As is true with most plants, these insects can be controlled with the use of beneficial insects. However, it is important that the pests be detected early and control measures be implemented before the insects get out of control. The use of insect barriers and yellow sticky traps for early detection and control is important. Installing an inexpensive misting system can help with humidity and temperature control and will also provide an added benefit of slowing the rapid outbreak of spidermites and other insects.

Sanitation is very important. Keeping the greenhouse free of debris and old plant material will help prevent the spread of fungus and insect problems. Keeping the outside of your building free from weeds and other insect attracting material, especially around your intake vents, will aid in preventing a rapid invasion of insect pests coming in from the outside

HARVESTING

The harvest of cucumber fruit usually begins about 30 days after transplant. The fruit grows very rapidly. A flower will yield a 1-pound fruit in about 10 days. The weigh gain of the fruit can be very rapid during the high light times of the year. A 3/4-pound fruit today will be a 1-pound fruit tomorrow. Fruit is usually harvested when it weighs about 1 pound. Harvesting may be required every day during high light periods. Fruit allowed to remain on the plant can grow to an unmarketable size of 3 - 4 pounds and will sap the energy of the plant.

When harvesting the fruit, it is important to remove the fruit stem back to the main stem of the plant. Any "stub" that is left can be attacked by the fungus botrytis. Harvesting early in the morning, while the fruit is still cool, will help maintain crisp fruit. The fruit is very thin skinned and should be quickly wrapped in stretch wrap or shrink-wrap film to prevent moisture loss. Once the cucumbers have been wrapped, they can be stored at 50When harvesting the fruit, it is important to remove the fruit stem back to the main stem of the plant. Any "stub" that is left can be attacked by the fungus botrytis.

Harvesting early in the morning, while the fruit is still cool, will help maintain crisp fruit. The fruit is very thin skinned and should be quickly wrapped in stretch wrap or shrink-wrap film to prevent moisture loss. Once the cucumbers have been wrapped, they can be stored at 50o - 55- 55o degrees F. for 2 - 3 weeks. You should avoid storing them with ethrel producing fruits and vegetables, because this will accelerate their ripening process and result in a shorter shelf life.

WRAPPING MACHINES

There are 3 different types of machines commonly used for wrapping cucumber fruit. The size machine used by a grower is dependent on the volume of production. The smallest is a stretch wrap machine. These machines are used in the meat and "deli" departments of most grocery stores. A grower can usually wrap 3 - 4 fruit per minute with one of these machines.

A mid-sized wrapping machine utilizes an L-sealer and a heat tunnel to shrink the film around the cucumbers. These machines are widely used in industry to "shrink wrap" a variety of products. With an L-sealer and heat tunnel, a grower can expect to wrap 15 - 20 fruit per minute.

A large cucumber grower may choose to use a special machine that is designed especially for wrapping cucumbers. These machines utilise 2 rolls of shrink film (one layer below the fruit & one layer above the fruit). These machines are capable of wrapping 100 - 160 fruit per minute

Target Production Estimates

Cucumber plants are capable of producing yields of 3+ pounds of fruit per plant per week during the high light times of the year. This is about 12 pounds per plant per month and across the normal 90-day harvest period, yields of 30 - 35 pounds per plant should be the target production goal of a grower. During the low light times of the year, production will decline. Your target production goals may drop to 1 - 1.5 pounds per plant per week

However, a grower who raises cucumbers on a year around basis (usually 3 crops), should target his annualised production at 60 - 90 pounds per plant space per year. This could also be stated as an annualised production of 7 - 10 pounds of fruit per square foot of greenhouse per year

MARKETING

Cucumbers are eaten - skin and all - there is no pealing required because the skin is not bitter. Unlike field cucumbers, the fruit has not been waxed, therefore there is no waste to throw away. The fruit is very low in calories.

The seeds do not mature, therefore they are considered "burpless". Once most customers have tried these cucumbers, they never want to return to using "field grown" varieties.

In the U.S., cucumbers are marketed in different ways. Along the East coast, cucumbers are sold in 12-pound boxes. In the Central and Western part of the country, cucumbers are sold in 16-pound boxes. These boxes are further defined by the size of fruit within the box. The most popular size is 1-pound fruit, therefore a 16-pound box with a 16 count of fruit. However, some boxes may be graded as 12 count (very large fruit), 14 count, and some boxes containing smaller fruit may be graded as 18 count.

All fruit is expected to be a dark green color and free of blemishes. They are expected to be relatively straight and evenly filled from end to end. If a slightly curved fruit is placed on a flat surface, the arch should not leave a clearance of more than 1" at the highest point.

Some large producers ship nationwide. There are large growers in Florida, California, Mexico, British Columbia and Ontario in Canada. The majority of these growers begin their heavy production during March & April and there is a secondary peak in production during October & November of each year. This is usually when the selling price of cucumbers goes down. Many of these growers are out of production during the Winter months and again during the Summer months. With less available supply, the market price is high during the Winter months with a secondary peak during the Summer months. The selling price usually ranges from $1.99 - $2.99 per pound during the Winter months to a low of about $.99 per pound during the Spring and Fall. Wholesale prices are typically 33% - 40% less than the retail prices

Local Grower

A local grower always has the advantage of providing

"fresh picked" quality. However, Cucumbers are a relatively unknown commodity. Many people will confuse these cucumbers with zucchini or some other strange exotic fruit. It is the grower's responsibility (and opportunity) to develop their own markets.

One method of developing a market that has been very successful for several growers has been to set up a "demo table" in the produce section of local supermarkets. Most produce managers are very receptive to this because their section of the supermarket rarely gets the opportunity to give "demo's". If the grower can arrange to spend just 2 hours per week doing "demo's" at the local supermarkets, they will probably develop more demand for their product than they can produce.

A "demo table" usually consists of a card table, a cutting board, cucumbers and toothpicks. The grower provides "free sample" slices for very everyone to taste. Many growers will also put together a poster board of pictures and descriptions of how and where the cucumbers are grown. Most growers will also label their fruit and ask the customers to look for their label when making purchases in the future.

Hybrid Cultivar Types

Early cucumber cultivars were populations or inbred lines produced commercially by open pollination of bulk-increased or hand-pollinated breeder seeds. In the 1960s, large-scale production of hybrid seed began as the result of development of gynoecious inbreds, and the availability of growth regulators to control sex expression. Hybrids have now taken over most of the commercial production in North America.

Most cucumber cultivars are hybrids produced by crossing one gynoecious inbred as the female parent with one monoecious inbred as the male parent. Greenhouse cucumber hybrids are usually produced using two gynoecious inbreds to produce a hybrid with very high gynoecious sex expression. Hybrids for home garden use are often produced using two monoecious inbreds to produce a hybrid with monoecious sex expression. Monoecious hybrids have the same total yield as gynoecious hybrids, but they have lower early yield, so require more harvests. For the home gardener, it is often preferable to have fewer fruit produced over a longer period of time. Three-way hybrids of pickling types were produced for a while during the 1980s, and involved crossing a gynoecious with a hermaphroditic inbred, and then crossing the hybrid with a monoecious inbred.

Inbred Production

Inbred seeds are produced in the inbred parent by pollination of their pistillate flowers using pollen transferred from staminate flowers on the same plant (self-pollination), or an adjacent plant (sib-pollination). The pollenizer (pollen vector) for cucumber is usually honey bees, but bumble bees and other insects will also move pollen from flower to flower. The inbred parent is planted in multiple rows in an isolation block or insect-proof cage. Isolation blocks are fields that are located at least one mile from the nearest cucumber field. Insect-proof cages will have one hive of bees added during flowering time for the cucumbers to be pollinated.

If the inbred line being increased is monoecious, there will be staminate and pistillate flowers produced on each plant. These will be sufficient for seed production. If the inbred line being produced is gynoecious, the plants will have to be treated with a growth regulator soon after seedling emergence to make them phenotypically monoecious. This is usually done by spraying the plants with a growth regulator (hormone).

Common growth regulators used to make gynoecious cucumber inbreds monoecious are gibberellic acid 3 (GA3), silver nitrate, and amino ethoxyvinyl glycine (AVG). Plants should be treated at the cotyledon or first true leaf stage, and one a week for up to four weeks (10-leaf stage). Rates used are 250 ppm of silver nitrate, or 250 to 500 ppm of AVG. Some inbreds, such as WI 2757, are highly gynoecious and are difficult to convert to monoecious sex expression. Those will require higher rates and more applications of growth regulators.

Hybrid Production

Hybrid seeds are produced in the seed parent by pollination of their pistillate flowers using pollen transferred from staminate flowers in the male parent. The pollenizer (pollen vector) for cucumber is usually honey bees, but bumble bees and other insects will also move pollen from flower to flower. The two inbred parents are planted in an isolation block or insect-proof cage in alternating rows, with 1 or 2 rows of the female parent for each 1 row of the male parent.

About six weeks after fruit set, or 12 weeks after planting, the cucumber fruit will turn from green to white, yellow or orange. The fruit will become softer and a few will have started to rot. At that time, fruit will be harvested from the rows containing the female parent. Rows containing the male parent will not be harvested, since they will be the male parent inbred, not the hybrid. Fruit will be put into a bulk seed extractor that crushes the fruit between rollers, and separates the fruit pieces from the seeds. Yield will be about 200 seeds per fruit, with an average of 427 lb/ A (700 lb/A from a good field).

In order to reduce the possibility of disease for the surrounding crops, and to prevent theft of the seeds of the cucumber inbreds, the field should be disked as soon as possible after the seeds have been harvested.

Commercial Seed Production

Most commercial cucumber seed production is located in semi-arid areas of the west such as California. Arid conditions favor the production of high quality, disease-free seeds. Sanitation is important at all stages of production. Workers should wash their hands with antibacterial soap or rinse them with 70% isopropyl alcohol before handling plants or fruit and between seed lots. All equipment should be cleaned and all soil and plant material removed before use in production areas. Clean and disinfect harvesting tools and equipment with alcohol or 0.5% NaOCl or Ca(OCl)2 between seed lots. Sanitation, harvest, and control procedures for production of foundation and stock (parent) seed should be at least as stringent as that for commercial seed.

The process of growing cucumber seed crops is similar to that for growing market crops except that site selection is more critical. Choose a field that has not had any cucurbits (watermelon, cantaloupe, honeydew, cucumber, summer or winter squash, pumpkin, or gourd) in it for at least 2, but preferably 4 years. A field that has a history of foliar diseases such as anthracnose should be avoided. Fields for open-pollinated cucumber seed production should be isolated by at least 1 mile from other cucumber fields to prevent contamination by outcrossing.

Selection of parental seed from elite or foundation seed is the first critical element of seed production. Use seed that was produced in dry climates and has been tested to be free of the pathogens causing gummy stem blight, anthracnose, and angular leafspot. Direct-seeded plantings are the standard method, and reduce the risk of seedling contamination in greenhouses.

Drip or furrow irrigation should be used in the production field instead of overhead irrigation to reduce leaf wetting and disease spread. Roguing of off-type and diseased plants within the field should be done throughout the growing season. There are four useful stages for roguing. The first is before flowering when vegetative characters are checked. The second stage is at early flowering when morphology of undeveloped fruit is checked.

The third stage is when the developing fruit are checked for trueness to type, and the final roguing is confirming the external morphological characters of the fruit to be harvested. Roguing for off-types is not effective after pollination in a field for open-pollinated seed production. It is only effective when fruit have been self or cross-pollinated and the male has no off-types. Inspectors should be trained to recognize common mixups.

All fruit should be inspected by trained technicians for symptoms that are suspected to be fruit blotch. All fruit suspected of having fruit blotch must be discarded. No fruit should be harvested from vines that have anthracnose or gummy stem blight symptoms. When seeds of fruit are mature the fruit are picked up into seed harvesters that crush the fruit and separate the seeds and pulp from larger fruit chunks. The seed slurry is transferred to bins where it is allowed to ferment for 24 to 48 hours. During this time the sugars and gelatinous material surrounding the seeds are degraded.

After fermentation, seeds are separated from pulp and juice by washing in a rotary washer or flume system. Seeds are dried using forced air warmed by propane heaters. Seeds are placed on flat drying beds or in large rotary dryers. Dry seeds are run through a mill containing sizing screens that separates large seeds from trash and small seeds. Seeds should be stored in hermetically sealed containers at low moisture content. Under favorable storage conditions, seeds should last 5 years. To be salable, germination of the seed lot must be at least 80%.

REFERENCES

Doijode, S. D., *Seed storage of horticultural crops*, Haworth Press, 2001.

Jarvis, W.R., *Cucumber Diseases*, Agriculture Canada publication, 1992.

Motes, J. E., "Pickling Cucumbers", *Production & Harvesting*, M.S.U. Cooperative Ext. Service Bulletin, 1977.

O'Sullivan, J., *Growing Pickling Cucumbers*, O.M.A.F. Factsheet, 1980.

12

Seed Potato Production

Potato is the term which applies either to the starchy, tuberous root vegetable crop from the various subspecies of the perennial plant *Solanum tuberosum* of the Solanaceae, or nightshade, family, or to the plant itself. In the region of the Andes, the word is also used to refer to other closely-related species of the genus *Solanum*. Potato is the world's most widely grown tuber crop, and the fourth largest food crop in terms of fresh produce — after rice, wheat, and maize (corn).

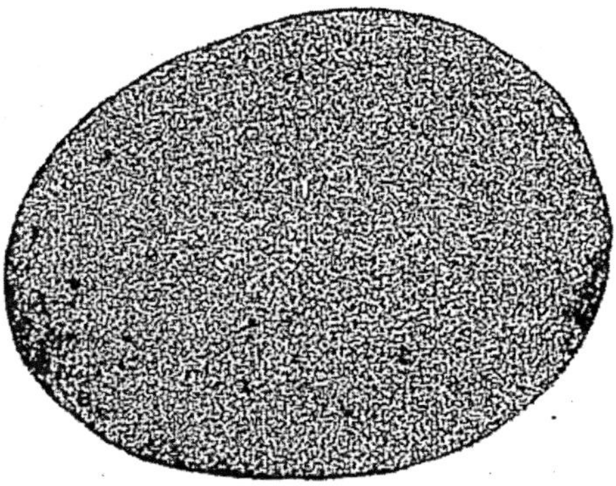

The potato originated in the area of contemporary Peru and Bolivia, identified more specifically in research published by David Spooner in 2005 as an area of southern Peru, just north of Lake Titicaca. The potato was introduced to Europe around 1700, and subsequently by European mariners to territories and ports throughout the world. Thousands of varieties persist in the Andes, where over 100 varieties might be found in a single valley, and a dozen or more might be maintained by a single agricultural household.

However, today, over 99% of all cultivated potato varieties worldwide are descendants of a subspecies indigenous to Chile.

DISEASE CONTROL

A variety of methods are used to monitor disease levels in the nuclear stocks, and to eliminate low levels of diseases if they become re-established. Test plants (plants that react in a specific way to infection by a specific pathogen) may be used to check for diseases. Gomphrena globosa is used to detect potato virus X; Rutgers tomato is used to detect potato spindle tuber viroid; Chenopodium quinoa is used to detect potato virus S; and Nicotiana glutinosa is used to detect potato viruses X, S and Y. Serological methods such as latex agglutination, slide agglutination, and ELISA (Enzyme-Linked Immuno-Sorbent Assay) are used to detect a number of the virus diseases. Culturing and staining methods are used to detect bacterial and fungal diseases. The PAGE (Polyacrylamide Gel Electrophoresis) test and Dot Blot test are used to detect potato spindle tuber viroid (PSTV).

At the nuclear stage, plants that test positive for any one of these diseases are discarded. Disease-free plants are further increased in the greenhouse by meristem culturing or rooted-cutting techniques. In Alberta, the initial increase is by meristem culturing. Tubers from these greenhouse-grown plants are then increased by seed potato growers in Alberta.

Diseases such as potato virus X (PVX), potato virus S (PVS), potato spindle tuber viroid (PSTV), blackleg, bacterial ring rot (BRR), fusarium wilt, verticillium wilt and potato leaf roll virus (PLRV) can re-infect seed potato stocks within one or two years. To prevent or slow down this reinfection process, Elite seed stocks must be grown under strict sanitary conditions.

SANITATION

Without adequate sanitation, diseases such as bacterial ring rot and blackleg can infect seed stocks from contaminated equipment or storages. All seed growers must be aware of sanitation procedures and apply them rigorously.

Sanitation consists of cleaning and disinfecting all equipment, storages, tools and pallet boxes that come in contact with the potatoes. Since most disinfectants are inactivated by soil and plant debris, it is essential that this material be removed from the equipment by a thorough cleaning with a hot-water or steam cleaner prior to disinfection. Then the equipment is thoroughly soaked with a disinfectant such as: bleach (sodium hypochlorite 10%); quaternary ammonium (such as Roccal, Shur-Gain disinfectant, or Bardac 2210); or formaldehyde. Follow the manufacturers' recommendation regarding the safe use of these products because they are all potentially hazardous to humans.

Sanitation also includes disposal of cull potatoes, and roguing and removal of diseased plants from the field. Cull potatoes should be removed from the farm as soon as possible and fed to cattle or buried in a landfill site. When roguing, diseased plants must be removed from the field and disposed of. They can become a source of continuing infestation by the disease if left in the field.

WHOLE VERSUS CUT SEED

North America potato growers have cut seed for nearly 200 years and we can expect that they will continue to do so for many more years. However, seed growers can benefit from planting small whole seed tubers.

Planting whole seed tubers eliminates the need to plant the higher classes of seed in tuber units because every plant grows from a different whole seed piece. If one plant is diseased only that one plant has to be removed. This makes roguing simpler. If seed is planted in tuber units, a diseased plant must be removed along with neighbouring plants on the assumption that they may have the disease.

Small whole seed tubers eliminate expensive and time-consuming cutting. Because cutting is eliminated the chance of spreading diseases from tuber to tuber is greatly reduced. The trend in commercial and seed potato production is to plant seed cut from smaller seed tubers. Seed growers can supply smaller seed tubers to their buyers if they plant whole seed because "single drops" set more tubers per plant than cut seed. Because of the increased competition the tuber size of the resulting seed crop is reduced.

Other advantages of whole seed are that plants are more uniform in their growth and therefore result in more uniform tubers at harvest time. Uniformity is further enhanced because whole seed results in near-perfect stands and this in turn results in fewer oversized and misshapen tubers. All growers, especially seed growers, should look at the advantages of whole seed and assess the benefits of using them in their operation.

SEED PREPARATION

The first consideration when purchasing seed for recertification is disease and size quality. Buying from a grower with a good reputation will reduce the risk of receiving diseased seed and increase chances of obtaining good seed. Some seed potatoes from the United States can not be recertified in Canada.

Seed potatoes, whether purchased or produced by the seed grower, should be warmed to 10-15°C for two weeks prior to handling. During grading and transporting, seed potatoes should not to be bruised or stressed in any manner. Rots, if any, should be removed prior to cutting.

Cut seed should be treated with a fungicide and planted within a few hours of cutting. If not planted immediately, the seed pieces should be suberized as outlined below for precut seed.

Cutters must be kept clean, sharp and disinfected. Sharp knives result in clean cuts that suberize quickly and reduce decay. Seed cutters must be adjusted to produce uniform seed pieces.

Precut Seed

Precutting of seed gives the producers more flexibility at planting time. Seed potatoes can be precut up to four weeks prior to planting provided they are treated properly. Potatoes that are to be precut should be warmed to 10-15°C and 95 per cent relative humidity prior to cutting and maintained at 15-20°C, and high humidity, for two to four days after cutting to promote healing of cut surfaces. If the potatoes are to be kept longer, they are then cooled to 4°C for the duration of storage and warmed again prior to planting. Seed-piece treatment should not be necessary after suberisation. If seed treatment is required, it should be applied just prior to planting but not after cutting. Precutting seed must not be attempted if the environmental conditions described here cannot be met.

Seed-piece Treatments

Fungicidal treatments that prevent seed-piece decay and fungal seedling disease are recommended, particularly if seed pieces are planted in cool wet soil or if a cool wet spring is anticipated. Several fungal diseases can decay seed pieces resulting in missing or weakened plants.

PLANTING

Higher classes of seed are planted first so they can be top-killed early to reduce infection by virus-carrying aphids.

Cup-type planters reduce the risk of disease spread and therefore are better than pick-type planters for planting seed potatoes.

Tuber-unit Planting

Tuber-unit planting is a process where all seed pieces cut from a single tuber are planted consecutively in a row and are separated from seed pieces of another tuber. Tuber-unit planting is used to identify tuber-borne diseases, separate them from current-season infections and make roguing easier. (All plants originating from an infested tuber are usually infected; however, when diseases are introduced in the current season only one or two plants are infected.)

Tuber units are planted either by hand (in very small plots), or with a tuber-unit planter. A tuber-unit planter is a manually fed planter that allows tuber units to be planted consecutively with or without the extra space. To facilitate planting, seed tubers can be pre-cut part way through and then separated into seed pieces at the moment they are placed into the planter cups. For larger areas, where manual cutting and planting is not practical, whole tubers can be planted using cup-type or tuber-unit planters. When cutting seed tubers for tuber-unit planting the knife must be dipped into a disinfectant such as 10 per cent bleach or a quaternary ammonium solution after each tuber is cut.

Fertility

Fertilizer should be applied as recommended by a soil analysis report. Total nitrogen to a depth of 24 inches (30 cm) should be about 125 to 175 lb/ac (110-160 kg/ha); and phosphorus about 100 lb/ac (90 kg/ha). Although most Alberta soils are not deficient in potash, grower experience and research has shown that when levels are below 350 lb/ac (315 kg/ha), potassium should be applied.

Maintenance of Seed Potato Fields

Diseases in seed potatoes must be controlled so that certification at the planned class is assured. The presence of seed-borne diseases may require that fields be rogued to keepthe level of disease below the tolerances for that seed class. Although not always practical, or possible, growers should strive for complete freedom from disease in all classes of seed potatoes.

To reduce or prevent the spread of viral diseases in seed fields, sucking and chewing insects such as aphids and leafhoppers must be controlled with insecticides. Growers who are in high-risk areas should use a soil-applied insecticide at planting and foliar insecticides during the growing season to reduce the spread of viral diseases.

Disease plants may have to be removed from seed fields if the numbers exceed tolerances. Weeds must be controlled in seed potato fields to allow for proper inspection and roguing. For weed-control procedures and chemicals refer to the section on pest management.

Irrigation and Disease Spread

Irrigation water can spread some diseases, such as blackleg and wilts. Wheel-type and pivot-type sprinkler systems may spread blackleg, potato virus S, potato virus X and potato spindle tuber viroid. Precautions like disinfecting wheels should be taken when moving equipment from field to field or from plot to plot, to prevent the spread of these diseases.

Roguing

Roguing is the removal of diseased plants from a field. All diseased plants including the new tubers and seed-pieces should be removed from the field. After handling diseased plants, avoid handling healthy plants and disinfect hands and tools as soon as possible. Fields should be inspected for disease on a regular basis.

Seed growers may be required to rogue diseased plants from seed fields to meet disease-level requirements for a particular class of seed.

HARVEST MANAGEMENT

The performance of disease-free seed potatoes is often directly related to the physical health of the tuber. Healthy, sound tubers will perform better than bruised or damagedtubers because they may be less prone to seed-piece decay or soft rot. Cool-weather harvesting, excessive chain speeds, high drops and lumpy soil may result in tuber damage. If some bruising has occurred, a post-harvest treatment with Mertect will help protect the tubers from storage decay.

Prior to harvesting and loading, all equipment must be cleaned and disinfected. Chains and conveyors should be maintained, repaired and adjusted properly to prevent bruising and scraping. Potatoes should not be dropped more than 6 inches (15 cm) in any harvest or piling operation.

Potato vines are top killed at least 10 days prior to harvest with diquat (Reglone) to mature the tubers and make harvesting easier. Mature tubers have better skin set and are less affected by normal harvesting procedures.

Seed potatoes should be top-killed as soon as they reach a sufficient size. Generally Pre-Elite, Elite 1 and Elite 2 should be top-killed by August 20. All vines and leaves must be completely dead within 10-15 days of vine killing to prevent late-season spread of leaf roll. Fields are normally harvested 10 to 20 days after top-killing. A longer period in the ground after top-killing is undesirable as it may increase the incidence of rhizoctonia in the harvested seed.

STORAGE MANAGEMENT

Each class and variety of seed must be stored separately to prevent mixing. Sprout inhibitors must not be applied to seed potato fields or in storages that are to be used for seed potatoes. Residues will evaporate and re-establish in the new storage and subsequently may retard sprouting of seed in the future.

Thoroughly clean and disinfect storage facilities and all grading and handling equipment after planting is completed and again prior to harvest. Cure potatoes at 13-16°C and a relative humidity of 90-95 per cent for 8-15 days to heal wounds and bruises. Seed potatoes should be stored at a temperature of 3-5°C with a relative humidity of 90-95 per cent for long-term storage. Allow only authorised personnel into the storage and visitors should disinfect their footwear before entering.

GRADING

Remove all culls from the seed farm as soon as it is practical. Alberta seed growers normally sell their seed in bulk, which must be graded according to the seed regulations unless the buyer specifies otherwise. Seed regulations allow Grade A seed to range in tuber diameters from 45 mm to 70 mm for long varieties, and from 50 mm to 80 mm for round varieties. Seed lots with this size variation result in highly variable seed pieces which cause unacceptable hill-to-hill variations. Growers may get a copy of the new regulations from the Food Production and Inspection Branch Agriculture Canada, in Calgary (table 1).

Table 1. Size classes of seed potatoes.

Variety Shape	Grade A	Grade B
Long	45 to 70 mm	30 to 45 mm
Round	50 to 80 mm	30 to 50 mm

When commercial growers pick up seed potatoes at a seed farm, their trucks must be disinfected before entering the seed farm property. This should be done away from the seed farm especially if the truck has hauled commercial potatoes.

MARKETING

New seed potato growers need to consider both where and how they will market their seed. This information will assist them in deciding the quantities and cultivars to be grown. The Potato Growers of Alberta, Alberta Agriculture and the Alberta Seed Potato Growers Association co-operate in promoting the sale of seed potatoes to export markets, particularly in Washington, Oregon and California. Growers selling independently should contact the office of the Potato Growers of Alberta for marketing information, advice and contacts. The Alberta Seed Potato Growers Association publishes an annual directory that lists all seed growers.

REFERENCES

Hawkes, J.G., *The Potato: Evolution, Biodiversity & Genetic Resources*, Smithsonian Institution Press, Washington, D.C., 1990.

Salaman, Redcliffe N., *The History and Social Influence of the Potato*, Cambridge University Press, 1989.

Stevenson, W.R., Loria, R., Franc, G.D., and Weingartner, D.P., *Compendium of Potato Diseases*, 2nd ed, Amer. Phytopathological Society, St. Paul, MN, 2001.

13

Seed Production of Sweet Potato

The sweet potato (*Ipomoea batatas*), commonly called a yam in parts of the United States (although they are not actually yams), is a crop plant whose large, starchy, sweet tasting tuberous roots are an important root vegetable. The young leaves and shoots are sometimes eaten as greens. The sweet potato is only distantly related to the potato (*Solanum tuberosum*). It is even more distantly related to the true yam (*Dioscorea* species), which is native to Africa and Asia.

The genus *Ipomoea* that contains the sweet potato also includes several garden flowers called morning glories, though that term is not usually extended to *Ipomoea batatas*. Some cultivars of *Ipomoea batatas* are grown as ornamental plants.

This plant is a herbaceous perennial vine, bearing alternate heart-shaped or palmately lobed leaves and medium-sized sympetalous flowers. The edible tuberous root is long and tapered, with a smooth skin whose color ranges between red, purple, brown and white. Its flesh ranges from white through yellow, orange, and purple.

PLANT PRODUCTION

Seed Stock

High yields of good quality sweet potatoes do not just happen. They are usually the result of good practices throughout the production operation. As in other crops, the quality of the seed stock is a primary factor influencing quantity and quality of yield. Only diseasefree, certified roots should be planted. Roots 1.5 to 3 inches in diameter produce the largest number of vigorous plants.

Varietal characteristics tend to mutate or run out rapidly. Growers should purchase registered or certified seed stock each year or maintain a seed production program using procedures comparable to those required for production of registered or certified seed. Breeder's seed is produced each year from stock carefully selected by the plant breeder to maintain each variety's productivity, root conformation and freedom from disease. This stock produces foundation seed. Foundation seed is produced from breeder's seed stock to supply needs of registered and certified seed growers. It is carefully rogued to ensure freedom from diseases, maintain high yields and eliminate mutations. Registered seed is from the first crop produced from foundation seed stock. Registered seed and foundation seed are produced from cut plants.

Certified seed stock is from the second crop produced from foundation seed. Certified seed stock cannot be re-certified. The amount of seed stock required to produce enough plants for a given acreage is determined by the variety, the size of the roots, whether the roots are presprouted, and conditions at the time of harvesting, curing, storing and bedding.

For maximum early plant production, growers may presprout sweet potato seed roots. Warming for two to four weeks at temperatures from 75 degrees F to 85 degrees F, with a relative humidity of 90 percent and adequate ventilation, is optimal. The exact presprouting conditions used will depend on varietal characteristics, individual operations, preferences and experiences. Presprout until most of the seed stock has sprouts approximately ¼ inch long. Sprout development can be delayed, if necessary, by lowering the temperature in the presprouting room.

Types of Beds

Plastic-covered beds (used for early production) produce plants in seven to eight weeks. Pull plastic tight and fasten it when covering the bed, but be sure the plastic can be easily rolled up for ventilation. Use clear plastic film for the cover.

Open field beds produce plants in 9 to 10 weeks after bedding. This is an economical method of plant production for the grower who does not plan to set early. One disadvantage of open field beds is that they generally produce fewer pullings than plastic-covered beds. Georgia Jet is very sensitive to cold, wet soil and often fails to produce plants in open field beds where these conditions exist. All bed types should be constructed to allow adequate drainage.

Fertilisation of Beds

Fertilise beds with 1 pound of complete fertiliser (10-10-10 or 6-12-12) per square yard. Apply an additional ½ pound of 10-10-10 after each pulling.

Bedding

Ideally, treated roots should be placed in the bed so they are not in contact with each other. Bulk bedding, a method used by most commercial growers, has been successful when recommended disease control practices are followed. Cover roots with 2 inches of soil. Separate and bed roots according to size to help achieve proper covering and uniform sprouting.

Watering and Ventilation

Water the beds enough to keep the soil around the potatoes

slightly moist, but not wet. Keep covers in place until plants begin to emerge. To produce hardy plants and to prevent scorching, ventilate on warm days after the plants are up. Air temperature in beds should exceed 90 degrees F.

FIELD OR SITE SELECTION

Soil Characteristics

One of the most important steps in sweet potato production is selecting the right soil based on the following characteristics.

Soil Texture

Soils intermediate in texture are best. Production in south Georgia has been quite successful on sandy loams and fine sandy loams. Sandy loams in middle Georgia and silt loams and lighter clay loams in north Georgia are preferred. Light-textured soils generally produce sweet potato roots with smoother skins than dó heavytextured soils. The lighter soils may produce longer roots, however.

Native Fertility

Soils with high native fertility are desirable. Soils with lower native fertility can have excellent production if a well-planned fertiliser program is implemented.

Soil Drainage

Surface and internal drainage are important when selecting a field. Poor surface drainage can cause wet spots, and improper drainage can cause excessive erosion, both of which reduce yields. Poor internal drainage will also reduce yields; this is generally not a problem in the light-textured soils of the Piedmont and the Coastal Plain. Soils with poor internal drainage are characterised by high moisture content and poor aeration.

These conditions cause sweet potato roots to be large, misshapen, cracked and rough-skinned. Soils high in organic matter will usually produce rough sweet potatoes that are damaged by scurf.

Soil pH

The optimal pH range for maximum yields of highquality sweet potatoes is 5.8 to 6.0. If using lime to raise pH, apply dolomitic lime because it also contains magnesium.

Crop Rotation

A three- to five-year rotation program will reduce the potential for disease development. Do not follow legumes because the resulting excessive nitrogen causes cracks, reduces yields of No. 1 sweet potatoes, and can result in greater disease problems.

Land Preparation

Prepare land well in advance of planting time. Have the topsoil loose, well worked and friable. Turn the soil several weeks before setting out sweet potato slips or plants. To allow beds to settle and become firm, make final bed preparation several days before setting. Beds should be at least 8 inches high and as broad as row widths will permit. Narrow beds dry out quickly and restrict root development.

Soil infested with root-knot nematodes should be treated with a nematicide before transplanting. Contact your county Extension agent for further information on nematode control.

Fertilisation

The native fertility of most Georgia soils is low. Consequently, liberal applications of a complete fertiliser are necessary to produce top yields of most crops. Sweet potatoes are heavy feeders, so they require heavy application of complete fertilisers. Base fertiliser recommendations on soil tests. Requirements commonly range from 400 to 1,000 pounds in north Georgia and from 600 to 1,500 pounds in the Lower Coastal Plain. There is no satisfactory substitute for a soil test.

Phosphorus and Potassium Recommendations

Apply all the recommended phosphorus during or near transplanting. Apply approximately ½ pint of a starter solution consisting of 3 pounds of 10-34-0 (or a similar material) mixed in 50 gallons of water to each transplant. One hundred to 150 pounds per acre of popup fertiliser promotes earlier growth, particularly in cool or cold soils. A good pop-up fertiliser is similar to or equal to 10-34-0 and should be relatively high in phosphorus and low in potassium. For early growth stimulation, band pop-up fertiliser 2 to 3 inches to the side of the plants and 2 to 3 inches below the roots.

One-half to two-thirds of the potassium either should be applied in two bands, each 2 to 3 inches to the side and 2 to 3 inches below the level of the plant roots, or should be incorporated into the bed before transplanting. Research shows that broadcasting over the entire field is usually less effective than banding. An acceptable alternative to field broadcasting is the modified broadcast method, in which the preplant fertiliser containing nitrogen, potassium, and any recommended micronutrients is broadcast in the bed area only.

For example, on a bed 72 inches wide, a swath 60 to 72 inches wide of fertiliser is uniformly applied over the center of the bed. Incorporation by tilling will help reduce water and wind movement of the fertiliser and will place some fertiliser in the root zone. Apply the remainder of the recommended potassium when the vines are about 12 inches long by banding it in an area on both sides of the row just ahead of the developing root tips. Usually, no additional potassium side dressing is needed unless heavy leaching rains occur. If they do occur on sandy soils, an additional application of 30 to 40 pounds of potassium per acre may be required.

Nitrogen Recommendations

Typical Coastal Plain soils require 60 to 90 pounds of nitrogen (N) per acre. Extremely sandy soils may need additional nitrogen or an increased number of applications. Piedmont, Mountain and Limestone Valley soils usually require 50 to 80 pounds of nitrogen per acre for sweet potato production.

Seasonal nitrogen requirements will vary depending on rainfall, soil type, soil temperature, irrigation, plant population, and method and times of applications. For typical Coastal Plain soils, one-half to two-thirds of the recommended nitrogen either should be applied in two bands, each 2 to 3 inches to the side and 2 to 3 inches below the root level of the plants, or should be incorporated in the bed before transplanting. Research shows that broadcasting over an entire field is usually less effective than banding.

An acceptable alternative to field broadcasting is the modified broadcasting method described in the "Phosphorus and Potassium Recommendations" section. Incorporation by tilling will help reduce water and wind movement of the fertilizer and will place some fertiliser in the root zone. Apply the remaining recommended nitrogen when the vines are about 12 inches long by banding in an area on both sides of the row just ahead of the developing root tips. Usually, no additional nitrogen side dressing is needed unless heavy leaching rains occur. If they do occur in sandy soils, an additional nitrogen application of 30 to 40 pounds per acre may be needed.

Magnesium, Sulfur, Zinc, Manganese and Boron Recommendations

If the soil test indicates magnesium is low and if lime is

recommended, apply dolomitic limestone. If magnesium is low and lime is not recommended, apply 25 pounds of elemental magnesium per acre. Apply a minimum of 10 pounds of sulfur per acre, 1 pound of actual boron per acre and, if the soil test indicates zinc is low, apply 5 pounds of actual zinc per acre. If soil pH is greater than 5.6, apply 8 pounds of actual manganese per acre.

TRANSPLANT AND TRANSPLANTING

Planting Dates

Plant after the danger of frost has passed. Where irrigation is available, good yields may be obtained in plantings made as late as the last of June.

Transplanting Quality

Weak and spindly transplants may decrease yields significantly. Use only fresh, relatively large, vigorous slips or vine cuttings. "String" potatoes do not yield a satisfactory percentage of vigorous slips. Maintain as many healthy leaves as possible on the transplants. Healthy leaves are required for food synthesis, which hastens root development and early establishment of the plant. Removing the lower leaves during transplanting reduces yields as much as 20 percent. Transplants should be set deep, with at least three nodes (joints where leaves are attached to transplants) below ground level. Increasing planting depths up to five nodes has been shown to increase yields.

Vine Cuttings

Save seed stock from vine cuttings only. This will contribute more toward reducing disease problems than any other single practice. Plant bed cuttings must be cut above ground - never pulled and then cut. Sanitize the cutting instruments by washing them in a fresh solution of one part laundry bleach and 99 parts water, then cut upward, never letting the knife touch the soil. Vine cuttings produce smoother, more uniform roots, and potatoes keep better in storage if they are grown from vine cuttings.

Spacing

Space plants properly because spacing has a drastic effect on the ultimate size of the roots. Set plants 8 to 12 inches apart in rows 3 to 4 feet apart. Use 8-inch spacing for plants set early in the season and 12-inch spacing for mid-season or later plantings.

Starter Solutions

Water the plants with a starter solution such as 10-34-0 at transplanting. The high phosphorus content of the solution enhances early root development and growth, and the liquid settles the soil and eliminates air pockets. During the first week after transplanting, replant where transplants have died or were skipped. A good stand not only increases yield but also decreases the number of off-grade sweet potatoes.

CULTIVATION

Properly prepared beds are about 8 inches high at the time of transplanting and feeder roots soon occupy the entire bed. To prevent damage to these roots, cultivate with equipment that does not scrape or remove the soil from the bed. Disk hillers, or similar implements that throw soil to the bed, reduce root damage and raise bed height with each cultivation. Final bed height should be about 10 inches. There is slightly less damage to the ridge and vines if the rows are cultivated in the same direction each time.

Weeds

Yield losses to weeds are often less obvious than losses from insects and diseases because weeds do not have the definite symptoms that diseases and insects do. Weed competition for moisture and nutrients reduces plant vigor, food manufacturing surface and yield, so make every effort to control all weeds throughout the growing season. Cultivation has been the customary method of weed control, but in recent years herbicides have played a larger role. When used correctly, herbicides are effective in controlling weeds and reducing the number of cultivations. Frequent, shallow cultivations control weeds when herbicides are not used. Never apply a herbicide until you understand how to use it and are prepared to follow the directions on the label.

DISEASES

Sweet potatoes are subject to a number of diseases that cause heavy losses in the field and in storage. These diseases can be prevented or controlled by following recommended practices throughout the season. The most common sweet potato diseases in Georgia are scurf, black rot, wilt (stem rot), nematodes, soft rot, bed blight and internal cork. Scurf, black rot and wilt are controlled by the same practices.

Scurf produces a grayish-brown to black surface discoloration on the roots. These discolored areas can occur in all shapes and sizes with no definite outline. They can be uniform over the surface but are usually worse on the stem end. Generally, the sweet potato skin is not broken and the brown discoloration is only skin deep, so it can easily be scraped from the surface. Black rot causes a black lesion to develop on the base of young potato stems before the plants are removed from the bed. The foliage of affected transplants appears yellow and stunted. A dark to nearly black spot appears on the potato in the field or in storage. These spots begin as small, round areas, but they enlarge until they cover the entire potato. The surface of the diseased spots has a metallic luster, and the tissue just beneath the surface is greenish.

Wilt-infected vines have a slight off-color of a yellowish tinge, followed by puckering of the foliage and wilting of the plant. The older leaves drop off first, leaving yellow leaves near the vine tip. On affected young vines, many short stems grow at the center of the hill, giving the vine a rosette appearance. If the skin is peeled away from the diseased stem, the exposed tissue shows a dark discoloration. The discolored tissue extends down the stem into the roots. Wilt does not cause the stem to rot unless saprophytic organisms enter the diseased vines.

Nematodes cause above-ground parts of the plant to appear stunted, sickly and greenish-yellow. Feeder roots may be either galled or completely absent. Sweet potatoes produced by infected plants are pitted and cracked.

Bed blight (Southern blight) is caused by a soilborne fungus that attacks the stems of young plants at the soil line, causing a white mold to develop. Infected plants turn yellow and die, creating dead areas in the plant bed. Soft rot is produced by the same fungus that causes bread mold. It enters the sweet potato through breaks in the skin caused by careless handling. Decay usually begins at one end of the potato and progresses rapidly to the other end.

Initial symptoms are soft, watery, stringy conditions, followed by a mass of grayish-black mold covering the rotten area. After decay, moisture escapes and potatoes gradually become firm, hard and mummified. Bacterial soft rot can occur in the seed bed, field, or even after harvest. Infected plants usually have long black stem lesions, although potatoes on diseased vines can show various degrees of soft rot development. Good sanitation practice is the best method of preventing bacterial soft rot.

Internal cork forms dark brown to nearly black areas that become hard and gritty when the potato is cooked.

Cork symptoms, which are sometimes difficult to see at harvest, are much more visible after two or three months in storage.

Insects

Even when the best available control measures are used, few sweet potato crops entirely escape insect damage. This damage decreases the quality of sweet potatoes by marring appearance, providing entry points for decay organisms, causing waste when cooked, and sometimes causing objectionable tastes. Roots containing insects or their excrement are usually unfit for human consumption.

At least 18 species of insects feed on sweet potato roots. Among those causing the greatest damage are the sweet potato weevil, whitefringed beetle, wireworm larvae and flea beetles. Other beetles and caterpillars can cause damage in some years.

Sweet potato weevils are the most destructive insect pest of sweet potatoes; fortunately, they occur in only a few counties in the southern part of the state. Adult sweet potato weevils are antlike beetles about ¼ inch long. The head and wing covers are dark metallic blue and the thorax and legs are bright orange-red. The adult weevil feeds on all parts of the sweet potato plant but prefers the roots. Feeding scars on tubers consist of tiny shallow holes that usually occur in patches. The eggs, laid in vines and roots, hatch in about one week. The legless white grubs with brown heads begin to tunnel through potatoes, causing the potatoes to be too bitter for human consumption.

The grubs are white, legless and about ½ inch long when mature. The grubs feed on the underground parts of about 250 different plant hosts. Whitefringed beetle grubs in the soil must be detected before planting if control measures are to be applied. The grubs cannot be effectively controlled after a crop is planted.

Wireworm adults, dark brown and about ¼ inch long, live near the soil's surface under trash and leaves. The adults do not feed on sweet potatoes. The larvae, which hatch from eggs laid in the soil during the summer and early fall, injure sweet potatoes. The larvae are white, cream or yellowish gray with reddish brown heads and tails. They are smooth, shiny and relatively hard-bodied. When full grown, they are ½ to 1 inch long.

Larval feeding causes small, irregularly shaped holes in roots. The feeding sites are concentrated in growth cracks or randomly scattered over the root surfaces. A one- to two-year lapse between plantings in fields that have been planted to a cover crop, corn or pasture grass effectively lowers the wireworm population. Sweet potato flea beetle adults are pests of sweet potato foliage, but larval injury to roots causes greater concern.

16 Adult flea beetles are black, about / inch long and 1 characteristically hop when disturbed. They eat narrow channels in the upper surface of sweet potato leaves, creating a characteristic and easily recognised injury.

Laid in the soil, eggs hatch into slender, white 16 larvae. When fully grown, the larvae are about / inch 3 long. The larvae make small, winding tunnels just beneath the skin of sweet potato roots. The tunnels are nearly invisible at first, but soon darken. As the potato grows, the skin over the tunnels splits away, leaving shallow scars on the surface.

Although large populations of sweet potato flea beetles are often found in sweet potato plantings, they usually cause little damage to the harvested roots. Apparently, the larvae feed mostly on the fibrous root system of the plants or attack the edible roots only under certain conditions.

Other insects, such as beetles and caterpillars that chew on the leaves or stems and leafhoppers that withdraw juice from plants also cause damage to sweet potatoes. Contact your county Extension agent for more information on controlling these pests.

HARVESTING

Harvest for the early fresh market as soon as a sufficient number of roots have reached market size. Harvest for storage when the maximum number of No. 1 roots has been produced. Inspect the field regularly to determine when to harvest, making sure to harvest before frost to prevent damage.

Prevent chilling injury during cool weather by removing all harvested sweet potatoes from the field before nightfall. Prevent sunscald by removing roots from the sun immediately after harvest. Sunscald can occur during as little as 30 minutes in the sun.

Some harvesters require vines to be cut with a rotary mower so the vines will not interfere with the digging operation. The grower with only a few acres can use a turning plow (with a tip of a wing removed) or a three-point hitch-chain digger. Larger growers might use more complex harvesters, which require very little labor, do not require vines to be cut previously and deliver potatoes directly into containers. Adjust and operate harvesters to obtain the least amount of skinning and bruising. When possible, remove cuts, cracks and other culls from sweet potatoes to be stored. Use care in future handling, transporting and grading to keep damage to a minimum.

CONTAINERS

Container selection is important in the handling and the proper curing and storage of sweet potatoes. Research shows that larger containers cause less damage than smaller containers when they are used to store roots.

Several types of storage containers are on the market, and each is designed for a specific purpose. Appropriate containers can help maintain the quality of roots throughout curing and storage and ensure fast, efficient and safe handling from harvest to final disposition.

Pallets and pallet boxes increase efficiency in handling large volumes of roots. Crates can be stacked on pallets in truck beds in the field and moved with a forklift. Pallet boxes that contain 800 to 1,000 pounds of freshly dug roots can be used for processing and fresh market sweet potatoes. Roots for processing can be run into the boxes as field run.

Rectangular crates stack better and allow more efficient use of space in a storage house than do round or tub bottom bushel baskets. When properly stacked, rectangular or square crates distribute the weight to the strong points on the crate - the corners and ends -and prevent excessive weight on the roots. Stacking height should not exceed 10 crates (or 11 feet, 3 inches, including the pallet). Stacking higher than this can result in buckling or breaking of the lower containers. High humidity adds to the possibility of damage to the lower crates because the wood absorbs moisture and becomes soft.

CURING AND STORING

Successful storage of sweet potatoes begins in the field with good production practices. If poor quality roots are placed in storage, only poor quality roots will be taken out of storage.

Proper environmental conditions in the storage house will permit storage up to 12 months. Maintaining proper temperature, ventilation and relative humidity prevents extensive sprouting, shriveling and excessive weight reduction.

Proper storage depends on:

— maintaining a clean and sanitary storage house.
— harvesting in time to avoid low temperature injury to the roots.

— storing only sound and healthy roots.
— arranging and stacking crates properly.
— curing properly.
— maintaining the proper storage conditions after curing.

Either electric or gas heat is satisfactory for sweet potato storage houses if the systems are properly designed and installed. If you heat with forced air gas heaters, be sure you have uniform air circulation throughout the house. If proper circulation is not provided, hot spots can develop in front of the heaters and cause severe losses.

Release heat from the bottom of the house rather than from suspended heaters. Good heat distribution is much easier when the heat is released at or near the floor. Never let hot air blow directly on the potatoes. Stack the crates in the storage room so that good air circulation can be maintained. When potatoes are stored in rectangular crates, leave an air space between every fourth or fifth layer.

Sanitation

Thoroughly clean and disinfect the storage house before placing sweet potatoes in it. Remove all old or decayed potatoes and other debris and clean the floor and old crates before storing potatoes. Clean and make any necessary repairs to the false floor so that crates will stack properly.

Fumigate houses that are located in areas where sweet potato weevils are a problem before placing potatoes in storage.

Curing

Sweet potatoes are stored successfully only if they are cured properly at the beginning of the storage period. Even with extreme care in harvesting and handling, a wound often occurs at each end of the root as a natural result of digging. These wounds and other breaks in the skin create areas where disease-producing organisms can enter the root. Therefore, these wounds must be allowed to heal over as soon as possible. Wound healing occurs as a result of cork formation over the cut or bruised surfaces.

Begin curing as soon after harvest as possible, preferably on the day the roots are harvested. Curing involves three things: temperature, relative humidity and ventilation. Curing is a wound-healing process that proceeds best at a temperature of 80 degrees F to 90 degrees F and a relative humidity of 85 to 90 percent. The process takes five to seven days. Ventilation is also necessary to supply fresh air.

In addition to promoting the healing of wounds and bruises acquired during harvesting and handling, these environmental conditions are necessary for development of a protective cork layer over the entire root surface. As the cork layer develops, suberin, a waxy material produced by the root's outer cells, is deposited.

This cork layer and suberin act as a barrier to decaycausing microorganisms and to excessive moisture loss during storage. Proper ventilation provides fresh air and helps maintain the proper temperature and relative humidity.

Fresh air allows potato roots to function normally during curing and storage. Because the sweet potato root is alive, chemical changes continually take place during curing and storage. These changes cause a buildup of carbon dioxide in the storage house, resulting in stale air, which is harmful to the roots and must be removed.

Storing

Store sweet potatoes at temperatures of 55 degrees F to 60 degrees F. Maintain the relative humidity at 75 to 80 percent throughout the storage period so the roots will not lose excessive moisture. Ventilate the house during storage.

Some wound cork formation and a slow change of some starch to sugar may continue for a short time.

Sweet potatoes can be stored in the same room in which they were cured; however, it is very important that the curing room be cooled uniformly - but fairly rapidly - from the curing temperature (85° to 90°F) to the storage temperature (55° to 60°F). Allow no more than 10 days for this temperature change. Potatoes allowed to remain at the high curing temperature too long lose considerable dry weight and moisture.

GRADING AND MARKETING

Wash and carefully grade sweet potatoes for fresh market to meet size and grade requirements. Remove all cut, cracked, diseased and unattractive roots to make the pack as attractive as possible. Buyers often pay a premium on the same U.S. grade because of better grading and attractiveness of the product. Also, repeat sales and profits depend on quality and appearance. Markets for processing sweet potatoes are limited in Georgia, but some are sold for this purpose. Field-run sweet potatoes with culls removed are bought in some instances. In other cases, size restrictions are placed by the buyer, who usually prefers the smaller grade roots. Before growing sweet potatoes for processing, be sure to contact the potential buyer to determine specific requirements.

REFERENCES

Fielding, W.J, Van Crowder, L., "Sweet potato weevils in Jamaica: acceptable pest?", *Journal of Sustainable Agriculture*, 1995

Gibbs, I.H., "Surveys of the sweet potato weevil *Euscepes postfasciatus* in Barbados during 1989 and 1990", *Proceedings of the Eight Annual Conference*, Grand Barbados West Indies, 1990.

14

Seed Sources of Organic Vegetables

The organic vegetable industry is flourishing due to consumers preference for organically grown produce over traditionally produced vegetables. As a result, an increase in the variety and selection of many vegetables in retail supermarkets and restaurants throughout the United States has occurred recently. With the new regulation requiring organic seed sources for organically labeled vegetables, many organic growers are searching for certified organic seed. Smaller seed companies have produced the majority of organically produced seed to date. The commercial seed sector is starting to provide a more diverse selection of cultivars, yet there are still many hybrids that are not yet available.

Commercial vegetable seed production in the United States occurs mainly on the west coast in Oregon, Washington, Idaho, and California. The weather conditions in these areas have low humidity, excellent soils, and a warm, sunny, dry climate during summer which is favorable for organic seed production. These environmental conditions allow for a lower incidence of disease pest problems and a longer growing season, which are necessary for high quality seed production.

The Northeast and Midwest have also produced vegetable seeds in the past but now with the demand for organic seed increasing, seed producers in these regions have begun producing organic seed. These areas can be successful for producing seeds of many crops, even though the climate does not permit an extended growing season.

Selection of a seed crop that will thrive in the environmental conditions of the selected area is critical for achieving economic profit for the seed produced. Though Florida is not presently a prime area for seed production, okra, sweet corn, field corn, watermelon, and some other cucurbits were produced using traditional methods in the past. Commercial peanut seed production also continues to thrive in Florida. There are areas of Florida that have the potential to produce organic vegetable seeds, such as watermelon, muskmelon, and other cucurbits. Conditions in Florida are not conducive for growing longer season crops such as lettuce and carrots due to prolonged high temperatures and other environmental conditions.

Many organic growers want to grow open-pollinated as well as hybrid cultivars. To be considered a hybrid, the seed must be the offspring of two parents that differ in one or more heritable characteristics. For example, a green bean cultivar that has excellent flavor but is susceptible to a certain pest crossed with a different cultivar that is tolerant of the same pest, the two cultivars could be crossed and the new hybrid might have both excellent flavor and be resistant to the pest, provided the parents are homozygous. Seeds collected from a grown hybrid cultivar will not be true to type when replanted and thus can not be saved to plant next years crop.

One of the alternatives to hybrids is open-pollinated cultivars. These are produced by selfing or outcrossing to the same parental type. Other open-pollinated cultivars are referred to as heirloom cultivars because they have been grown for many decades. There is no standard length of time that a cultivar has to be grown in order to be considered an heirloom cultivar but generally 40 years is minimum. Since hybrid production requires obtaining the two inbred parents for crossing, they are mostly produced by commercial seed companies who own or develop the parental stocks. Thus, for smaller commercial seed companies and producers who grow their own organic seeds, only open pollinated cultivars are produced.

CONVENTIONAL AND ORGANIC SEED PRODUCTION

Vegetable crops are produced for two distinct markets, fresh and processed. Conventional cultural practices are fairly similar. Growing crops for seed requires a longer season since the crop must stay in the field twice as long as a conventional crop harvested for fruit and also for increased monitoring to ensure high seed quality and purity.

Standard production of vegetable seeds requires chemical herbicides, insecticides, fungicides, and fertilisers. An increase in the amount of chemical products used on seed crops may occur due to the length of time the crops remain in the field. As a result, plant diseases and insects have more time to attack the crop during seed maturation. With these factors playing a role in conventional seed production, the challenges for organic seed production are increased.

Organic farming practices began on small farms in the United States during the 1940s. As the diversity of crops labeled organic increased, there became a need for certification standards and regulations to ensure proper guidelines for growing organic crops. By the 1980s, the organic industry increased dramatically, resulting in the establishment of the Organic Certification Board which issued the Organic Foods Production Act (OFPA), identifying "organic" produce. The producer must use organically grown seeds, annual seedlings, and planting stock, Except, That, (1) Nonorganically produced, untreated seeds and planting stock may be used to produce an organic crop when an equivalent organically produced variety is not commercially available, Except, That, organically produced seed must be used for the production of edible sprouts.

LAND SELECTION

Good farming practices are especially important when growing vegetable crops for seed. Proper selection of land is critical to ensure maximum quality and economic yield. Crops should be rotated to reduce pest problems and any potential for seed contamination by open pollination with similar species types.

The seed production field should not have known weed problems that are too difficult to control through organic means as weeds will compete significantly with the crop, resulting in seed yield and purity problems. Weed management is a major problem with organic crop production.

Land Preparation and Soil Fertilisation

Soil should be tilled to ensure a fine seed bed which is critical for germination, particularly with small seeded crops. The soil should have good water holding capacity to allow for uniform germination and continued vegetative growth. In Florida, the beds should be raised and shaped depending on rainfall. It is especially important to have uniform seedbeds because the seeds are often precision planted and uniform emergence and seedling development are required for optimum management. Proper irrigation is necessary to ensure growth during droughts. Drip irrigation is an effective way to deliver water to the crop without wetting vegetative plant parts thus reducing potential disease problems.

It is important that the fertility of the soil is improved when producing organically since chemical fertilisers can not be used. To ensure good fertility and fewer soil diseases, crop rotation, use of a cover crop, green manure crops, mulch, animal compost, and plant material compost can all be used. Floridas soils are generally sandy which provides good drainage and ease of tillage, yet nutrients are leached readily during heavy rains. Organic or natural fertilisers are generally made from bone meal, blood meal, horse, and chicken manure, mushroom compost, ash from sunflower seed hulls, and many other substances. Considerations and modifications must be made depending on the specific land selection and soil conditions. There are many commercial organic fertilisers available for organic crops with the major ones listed below.

— Composted manure (mushroom compost is mostly manure), or blood meal or feather meal, used to increase nitrogen content. Dried blood is also a useful nitrogen source when used with wood chip mulching, which is high in carbon, and low in nitrogen.
— Wood ashes, or potash, for increased potassium.
— Rock phosphate, often crushed rock that contains elevated levels of phosphate.
— Seaweed extract - not a fertiliser, but aids plant growth and resistance to pests and disease.

Inoculates used with bean and pea seeds provide beneficial fungi which work with the plant's roots to help them fix nitrogen from the air. Cover crops used between seasons as green manure may increase soil organic matter for the growing season.

PLANTING TECHNIQUES

Seeds are generally planted directly by drilling in the field or transplanting from a greenhouse grown seedling. These seeds should be organically produced and all transplants must be organically produced. The seeds must be planted in such a way that proper vegetative development occurs that will support fruit and seed development with proper spacing and depth in the bed. Vegetable seeds are greatly influenced by temperature at planting; pea, lettuce, radish, beets, onion, and spinach grow best at cool temperatures for optimum emergence. Other crops including squash, melon, tomato, pepper, and eggplant grow well at warmer soil temperatures. In general, the best germination temperature for all these crops is around 70°F.

Row spacing and plant densities must allow for maximum plant development of the flower and unrestricted access to inflorescences for pollinators which will lead to proper fruit and seed set. Proper spacing will also allow for improved air movement reducing pathogens and space for harvest operations at the end of the season.

Pollination of crops for seed is also affected by row spacing and planting densities; using larger spacing allows for better wind movement and more room for insect pollinators. Crops that are wind pollinated include sweet corn and spinach and those that are insect pollinated include carrot, onion, and broccoli.

WEED, INSECT, AND FUNGAL MANAGEMENT

Management of weeds and pests are critical to ensure that organically produced seeds have high yield and quality. It is common for insect pests after harvest of the primary host crop to move to another crop in a neighboring field. This is the case for the Lygus bug. It will feed on legume crops but once harvested, they will move to umbiliferous crops such as carrots. Other insects such as cabbage looper infest crops in the Brassica family and causes loss of vegetative parts effectively reducing seed yield and quality. Aphids are a problem, especially at harvest because they secrete "honeydew" on crops including tomato, squash, and pumpkin making harvesting difficult. They also spread diseases, especially virus, which is difficult to control conventionally and even more challenging for an organic grower.

Because the seed crop is in the field for a longer period of time, there are more opportunities for multiple pathogens to interact with a single crop. There are also cases where pathogens target the seed directly and affect the entire seed population. To control these pathogens organically is complex and requires proper growing conditions such as temperature and humidity to minimise the incidence of such pathogens as powdery mildew which thrives in high temperature and humidity. Seed producing fields should not be located near commercial vegetable fields. Eliminating weeds will help decrease insect habitats and pathogen spread from host plants likely to infect the seed crop.

BIOLOGICAL PEST CONTROL

Another viable pest control method is through biological control. This practice is highly compatible with organic seed production, and it does not involve the use of chemicals. Biological control is defined as the decline of pest populations by natural enemies. Biological control utilises three sources of natural enemies that can be used to control harmful pests and reduce the use of organic pesticides. The first group includes parasitoid insects, which lay their eggs inside another insect. An example of a parasitoid insect is the wasp, Aphidius colemani which lays its eggs in aphid adults. The wasp matures inside the aphid egg, killing the aphid and continuing the wasps life cycle. The second group of beneficial insects is predators, which eat other insects. A common predator is the lady beetle, which preys on insects and mites.

The third group includes the weed feeders, which feed on weed populations. Insects such as a weevil, Hylobius transversovittatus feed on certain weeds including purple loosestrife. Pathogens such as fungi, viruses, bacteria, and nematodes are also capable of destroying weed populations and are called bioherbicides. Reducing weed populations is highly beneficial because often times during winter, pests live in weeds surrounding fields used for seed production. It is important to note that there are many beneficial insects that can be used to control a large number of pests.

HARVESTING, THRESHING AND DRYING

Vegetable seed crops are separated into two categories - those that produce wet-seeded fruits and those that are dry seeded. Wet-seeded crops include tomatoes, peppers, cucurbits and those that are dry seeded include the brassicas, legumes, and onions. Harvesting method depends on the type of seed being produced.

When harvesting dry-seeded crops, seed shattering must be prevented because seed harvest generally occurs after the crop reaches physiological maturity. To reduce shattering, the stalks of the plant need to be cut while still green and field dried allowing for uniform seed maturation. Once the seeds are mature, they can be machine or hand extracted. Observation may help to judge how the plant is maturing and for signs such as drying of florets or other plant parts. This can be complicated and it requires good knowledge of the crop but can lead to excellent seed quality, because the seeds will be harvested at peak maturity and size.

Determining when wet-seeded crops are mature for seed harvest depends primarily on fruit color. Peppers and tomatoes are harvested at full color because the percent germination is much higher in the mature red compared to the green fruit. In any case, wet-seeded fruits can be harvested, the pulp processed, and the seeds removed. The seeds must then be washed in water then dried either in the sun or by rotating in a large drum before oven drying. Modifications must be made with certain crops, such as melons, to remove the mucilaginous coating around the seed. When the seeds are removed from the pulp and outer coatings, they should be dried immediately to prevent fermentation.

CLEANING AND STORAGE

Once seeds are harvested, threshed, and extracted, they should be evaluated to determine if further cleaning is required to remove any debris such as dirt or plant parts. Small seeds such as carrots are sorted and separated using a debearder machine to remove unwanted outer seed parts. All seeds should be single units and all stored according to their individual temperature/humidity requirements. Seeds are best stored at 50% relative humidity at 50°F.

ORGANIC SEED TREATMENTS: PELLETING TECHNIQUES

The commercial seed industry often pellets seed, if the seed is irregularly shaped or small. Developments in seed technology have allowed for artificial seed pelleting materials to aid in regulation of water uptake by the seed which can influence germination rate. Seeds are pelleted with materials that coat with multiple layers of inert materials which shape and color the seed for better visualisation during seeding.

MARKETING

Commercial seed production starts with a breeder who develops a new variety. A portion of the original "breeder stock" always stays in the hands of the person who has developed that variety. Considered the purest form, breeder stock constitutes the "gold standard" for that variety, according to Dr. Jeff McCormick. A portion of the breeder stock becomes the parent of a larger quantity of foundation stock. The institution associated with the breeder controls the production of foundation stock, and in turn supervises production of registered seed for distribution to licensees, such as seed companies. These companies, in turn, contract (often with farmers) for a large quantity of certified seed. The final stage is production of seed from parent stock of certified (or select) seed for general distribution through commercial channels, although certified seed may be the final stage for large-scale grain production. Select is a term used more for vegetable seed, comparable to certified for grains.

For information on university foundation seed stock programs, see the Web sites of most land-grant universities. Seed companies routinely drop older varieties in favor of new ones (often hybridised, plant variety protected, and, sometimes, patented). This practice gave rise in the 1980s to grassroots efforts-epitomised by organisations like the Seed Savers Exchange (SSE)-to preserve older varieties through seed-saving networks. SSE organised backyard gardeners to raise and distribute seeds of heirloom vegetable crops that might be especially adapted to discrete geographical regions, might form part of the heritage of an indigenous (or other) population, and, most important of all, could be saved by the grower from year to year because they are "open-pollinated" (self-pollinated or vectored from another plant of the same type) rather than hybrid (produced artificially by controlled cross-breeding).

Commercial-scale organic production requires seed stocks (both open-pollinated and hybrid) with proven reliability-especially natural resistance to insects and diseases, as well as natural vigor to germinate promptly and out compete weeds. Good flavor and quality typically are considered more important than shippability. Additional attributes making for successful organic propagation are beginning to be identified.

Recently, organisations such as the Organic Seed Alliance (OSA) and the Public Seed Initiative (Cornell) have outlined a new public participatory model for breeding organic seeds. The model aims to strike a middle course between the inexperience of seed-saving farmers and any special-interest bias in formal research. Prior to training, farmers often lack the skills to select traits important for enhancing organic production. They may also lack resources to carry on multi-year development of seed lines. Leaving the research agenda in the hands of institutions simply accelerates the movement toward genomics and patentable outcomes.

Seeds of Change is leading the way in developing summer squash for organic production, especially zucchinis, emphasizing large canopies to shade out weeds, resistance to weather swings, adequate yields, and flavor. A preliminary evaluation of heirloom varieties at Cornell under organic conditions has identified a forgotten cantaloupe with superior flavor. 'Hannah's Choice' thrives under organic conditions, when grown for local markets and not for long-distance shipping.

Farmer Compensation

Exactly how farmers participating in breeding the new organic varieties will be compensated for their time is not clear, except that the farmers will ensure organic versions of their favorite regional varieties for their own use. Neither has anyone offered a clear distribution model for the new varieties. One possibility is the collaborative model. Plant breeding clubs share seeds among their own members, and the membership model has emerged as the preferred method for organic farmers to obtain transplants. The Organic Seed Alliance calls for "developing new relationships and exploring novel avenues of collaboration to bring quality seed to the organic movement." In the U.S., plant breeding clubs generally include a group of farmers assisted by a university researcher or other technical assistance provider.

Horizontal Selection and Breeding

Professional plant breeders have never focused on breeding for horizontal resistance, at least for the past 65 years.

During the 1960s, many plant breeders also begain to doubt the profitablity of breeding for vertical resistance (narrow selection for one or very few specific traits). The commercial life of most vertically resistant cultivars was too short to justify the amount of neessary work. The short market life of new introductions, combined with the development of improved crop protection chemicals and the financial involvement of chemical industries in breeding, led to abandonment of resistance breeding altogether, in favor of crop protection by chemicals. At present, the world spends about nine billion dollars annually on pesticides. Despite this, pre-harvest crop losses due to pests and diseases are estimated at 24 percent. In food crops alone, these losses are enough to feed about one billion people.

The only effective means of overcoming corporate and scientific opposition to horizontal resistance (broad selection for an array of resistance traits) is to make plant breeding as public and as widespread as possible. It is not necessary to find a good source of resistance, as when breeding for vertical resistance. Transgressive segregation within a population of susceptible plants will usually accumulate all the horizontal resistance needed. Should this not occur, merely widening the original genetic base will probably remedy the situation. Transgressive segregation, a common term in plant breeding, is "the segregation of individuals in the F2 or later generation of a cross that shows a more extreme development of a character than either parent gene."

A second step is the use of recurrent mass selection as a breeding method. Robinson originally recommended about ten to twenty original parents. Dr. Jeff McCormick, of Garden Medicinals and Culinaries, recommends fifty to one hundred, usually high-quality modern cultivars, but also some older landraces, for exposure to cross-pollination in all combinations. The progeny should total some thousands of individuals that are screened for resistance by being cultivated without any crop protection chemicals.

The majority of this early screening population dies, and the insect and disease pests do most of the work of screening. The survivors become the parents of the next generation. This process is repeated until the research group determines that enough horizontal resistance has accumulated. Usually 10 to 15 generations of recurrent mass selection will produce high levels of horizontal resistance to all locally important pests. The process could take ten to fifteen years in temperate climates, but less where more than one cycle per year could be realized. McCormick has recently streamlined the process suggested by Robinson in 1996 to about five generations.

Recurrent mass selection must be performed "on-site"-that is, in the area of future cultivation, at the time of year of future cultivation, and according to the future farming system (i.e., organic production). This will produce new cultivars that are in balance with the local agro-ecosystem.

ISSUES WITH THE CONVENTIONAL SEED INDUSTRY

Heretofore, the increasingly consolidated seed industry has served as the main engine of commercialisation and distribution of new introductions by producing certified (for grain crops) and registered (for vegetable varieties) seed. The industry has sought greater returns for its crucial service by acquiring intellectual property rights to seeds of unique varieties, limiting the number of varieties sold, and most significantly, finding advantageous legal or legislative avenues. A main attraction of biotechnology for seed companies is enhanced worldwide market share, not improved yields (as the case of Bt corn has shown). Accordingly, Gunnar Rundgren, president of the International Federation of Organic Agriculture Movements (IFOAM)-concurring with the assessment of World-Watch Institute-asserts that in the case of GMOs (genetically modified organisms) there are no benefits for either consumers or producers- only for the companies producing and selling them. If farmers feel they need herbicide-resistant varieties, that is because they are locked into a production system that depends on chemical inputs...[a system] that leads to further degradation of the environment, increased dependency of farmers and more risks for everybody.

Acquisition of exclusive ownership of seed varieties is limited under the 1970 Plant Variety Protection Act, which safeguards the rights of farmers and gardeners to use their own saved seed, and the rights of plant breeders to use PVP varieties for breeding new varieties, while affording seed developers a means to recoup their investment. Lobbying groups demanded protection for small farmers in the PVPA legislation. Seedsaving farmers and gardeners had become concerned by the European ban on many traditional open-pollinated varieties as part of a program of varietal "standardisation."

However, under an obscure 2001 U.S. Supreme Court decision (Pioneer Hi-Bred International vs. J.E.M. Ag Supply), companies for the first time could freely patent plant varieties under the 1795 U.S. Utility Patent law, without any reservations to protect small growers or farmers who wished to save (and sometimes sell) seed from their own crops. So far, this has affected mainly U.S. commodity grain crops. At the end of 2004, owners of patents on genetically engineered varieties had filed 90 lawsuits, involving 147 farmers and 39 small businesses, alleging seed patent violations.

ISSUES IN ORGANIC SEED SOURCING FOR COMMERCIAL GROWERS

In setting as a key goal for the future of public breeding, "development of 'a road map for invigorating public domain plant and animal breeding to meet the needs of a more sustainable agriculture,'" the 2003 Seed Summit committed itself to the totally new area of breeding for organic production. In doing so, it shifted ground beyond increasing the supply of currently available varieties of organic seed to developing new varieties designed specifically for organic production.

Industry Positions on Testing for GMOs

A system of tolerances for GMO contamination may eventually need to be established for certified organic crops-especially wind-pollinated crops like some grains and oil-seeds. Governmental agreements, especially on harmonisation of organic standards, would open the door for U.S. organic farmers to participate in foreign trade. Other suggestions include setting aside areas of the world still remote enough to produce foundation stock of wind-pollinated crops or establishing a U.S. government public seed bank of pure stock.

Quality Issues in Farmer-saved and -traded Seed vs. Purchased Commercial Seed

The highest quality grain seed sold to farmers is "certified," with minimum standards for purity, germination, test weight, trueness to type, and absence of physical damage. Ideally, seed for planting organic grain crops would be both "certified" and "certified organic." Shortages of certified organic grain seed have sometimes led farmers to use "bin-run" seed from a nearby organic farm or from a previous year's harvest that (while it is "certified organic") may contain light or broken seed, weed seed and other foreign matter, or pathogens. Such seed is also likely to germinate poorly. This is not invariably the case, of course.

Value in Going Back to Certified Seed Every Few Years if You Save Your Own

Although Canadian farmer Percy Schmeiser asserts that he selected and saved seed most of his 35 years of growing canola crops-thereby developing a landrace adapted to Saskatchewan conditions-the unfavorable outcome of the internationally publicised court case in which he was involved with Monsanto underscores the advisability of commercial farmers going back every few years to a reliable source of organic seed of their preferred variety. This practice guards against disease buildup, inadvertent contamination of the stock, and reversion of the crop to undesirable traits. This reliable source can be certified seed from a conservator university or commercial seed company. Jeff McCormick, a pioneer new-breed seed company owner, has suggested that vegetable farmers growing a contract seed crop may find it to their advantage to go back to the company every year for "select" (certified) seed for the vegetables they are raising for market, as well.

GLOBAL OUTLOOK

Another major issue at the World Seed Conference was intellectual property rights, or the implications of governmentally approved lists of permitted varieties. This is a special concern for traditional farmers in many countries, who are used to saving seed from year to year and have over the centuries developed unique landraces. A recent example is in Iraq, where a new report by GRAIN and Focus on the Global South cites a U.S. edict in occupied Iraq that "prevents farmers from saving their seeds and effectively hands over the seed market to transnational corporations."

Traditional practices of indigenous farmers are mostly compatible with organic production: planting a mix of adapted types (landraces) to ensure some survivors, despite vagaries of weather and insect/disease attacks; use of older varieties geared to minimizing capital investment; hand-harvesting and other labor-intensive practices precluded by modern, uniform, machine-harvestable varieties; and use of labor-intensive crop protection strategies like hand weeding and watering, rather than purchased off-farm inputs.

But can hand labor feed burgeoning urban populations, or is it a relic of a younger, less densely populated Earth, where 98% of people grew their own food? In the best of all possible worlds, a blend of traits uniquely adapted to organic production (not only resistance to local pests and diseases, but improved vigor and flavor) will result from horizontal breeding. This implies a far more decentralised food production system than we have at present.

Geography of organic seed production has ramifications mainly in the context of GMOs. Spain and Italy raise seed for the rest of Europe. Traditionally U.S. garden seed has been produced in Idaho and other arid West Coast and Intermountain regions. Relative severity of pest and disease pressures is a major consideration in producing quality seed. However, labor costs for seed production became an issue in the 1980s, leading to seed production for commercial growers as far away as Taiwan and Argentina-a development worrisome on several counts, not the least of which is the newly announced Chinese plan to invest billions of dollars in Argentina and Brazil in return for access to land and natural resources, an agreement finalised at the recently concluded Summit in Chile. Argentina has been identified as an emerging leader in GMO crop production.

REFERENCES

Hawthorn, L.R., Pollard, L., *Vegetable and Flower Seed Production*, The Blackiston Company, New York, 1954.

McDonald, M., Copeland, L., *Seed Production Principles and Practices*, International Thomson Publishing, 1997.

Peet, M., "Practices and Materials Permitted in Some Organic Certification Programs", 2002.

White, J.M., *Organic Vegetable Production*, UF/IFAS Coop. Ext. Serv., HS720, 1995.

15

Seed Systems and Security

To strengthen seed systems, we need to have a comprehensive understanding of them. Farmers, particularly small farmers, are involved in multiple kinds of seed systems, which help them produce and obtain the seed they need. These systems can be broadly divided into two types: a formal seed system and a local system. The local system is also sometimes called the "informal," "traditional," or "farmer" seed system.

FORMAL AND LOCAL SEED SYSTEMS

The formal seed system is the easier to characterize, as it is a deliberately constructed system that involves a chain of activities leading to clear products: certified seed of verified varieties. The chain usually starts with plant breeding and selection, resulting in different types of varieties, including hybrids, and promotes materials leading to formal variety release and maintenance. Guiding principles in the formal system are to maintain varietal identity and purity and to produce seed of optimal physical, physiological and sanitary quality. Certified seed marketing and distribution take place through a limited number of officially recognised seed outlets, usually for financial sale. The central premise of the formal system is that there is a clear distinction between "seed" and "grain." This distinction is less clear in the local, farmer seed system.

A local seed system is basically what the formal system is not. Activities tend to be integrated and locally organised, and the local system embraces most of the other ways in which farmers themselves produce, disseminate, and access seed: directly from their own harvest; through exchange and barter among friends, neighbours, and relatives; and through local grain markets. Encompassing a wider range of seed system variations, what characterises the local system most is its flexibility. Varieties may be landraces or mixed races and may be heterogeneous (modified through breeding and use).

In addition, the seed is of variable quality (of different purity, and physical and physiological quality). The same general steps or processes take place in the local system as in the formal sector (variety choice, variety testing, introduction, seed multiplication, selection, dissemination and storage) but they take place as integral parts of farmers' production systems rather than as discrete activities. While some farmers treat "seed" as special, there is not always necessarily a distinction between "seed" and "grain." The steps do not flow in a linear sequence and they are not monitored or controlled by government policies and regulations. Rather, they are guided by local technical knowledge and standards and by local social structures and norms.

Despite, or perhaps because of, their variability and local specificity to needs and preferences, local channels (e.g. household stocks, markets and social exchange networks) provide most of the seed that most small farmers use. Common figures suggest that somewhere between 80 and 90 percent of the seed that farmers access comes from the local seed system, although this varies greatly by crop and region.

In studies of seed systems, much has been made of the notion that small farmers, especially in vulnerable regions, strive at all costs to save their own seed, and that they get the bulk of what they sow from previous harvests. While this is broadly true, especially in remote or marginal areas, studies that actually quantify seed-source use find increasingly that, within the local seed system, local grain markets are also crucial in meeting seed needs, especially for poor farmers and in difficult times. Again, this varies greatly by crop. For many farmers, local markets are the second-best bet (after home stocks) as they put on offer the same varieties that farmers routinely sow. One study in southern Somalia demonstrates that where grain traders invest in obtaining good-quality "seed" (making a distinction from bulk grain), local markets can be a preferred source of replacement seed.

SEED SYSTEM INTERFACES

Figure 1 reproduces one of the most cited and useful figures illustrating seed system components. It shows formal and local seed systems as distinct but intersecting.

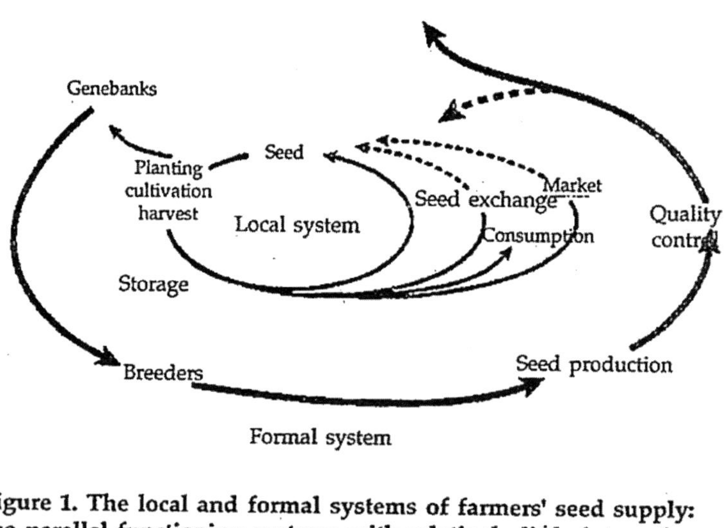

Figure 1. The local and formal systems of farmers' seed supply: two parallel functioning systems with relatively little interaction

It is perhaps this melding of the formal and the local that is of prime interest for those striving to create stable, resilient, and dynamic seed systems - systems on which farmers can actually rely. Three aspects of the intersections between the formal and local systems are of note.

— Materials flow between the two systems and create something new and potentially useful. Varietal identities frequently become blurred through several processes. For example, many successful so-called "modern varieties" from the formal system may be pure line selections from local landraces or may be incorporated as components of farmers' mixtures. Furthermore, farmers may recognise that modern varieties cross with local materials, such as criollo maize in Mexico. Farmers may also come to consider the modern introductions as "local" and eventually group them within the local genetic heritage. All these processes, frequent and ongoing, serve to muddle the divide between what is new and old, what is pure or not and what is local and what is modern.

— Farmers often draw seed from both systems for different kinds of crops. What is sometimes called the "maize craziness" of the southern African region serves to show how the same small farmers may routinely access seed from very different seed channels. The maize hybrids, sometimes renewed yearly, have to be accessed, directly or indirectly, through formal seed systems (stockists, commercial cooperatives, government parastatals and, sometimes, relief aid). Beans, however, may come from the farmer's own harvest or local grain markets, and the local sorghum seed may be passed from neighbour to neighbour. In some societies, in contrast, vegetatively propagated crops, like banana, never move through formal channels, and even hinting that they might be sold or bartered for profit may constitute a local social offence.

— Farmers sometimes use different channels for the same crop. It is not unusual for a farming household to meet their single-crop needs from varied seed channels. For instance, well-documented studies show that, during a single season, bean farmers throughout eastern and central Africa obtain some of their bean seed from their own stocks, some from markets and some from neighbours, and may obtain a handful of new material (to try) from an extension agent or research station. Broad strategies also frequently vary according to the farmers' wealth.

Some of the key attributes that need to be monitored and addressed in assessments of seed systems are indicated below.

They are complex and dynamic, but the principal unifying trends can usually be understood relatively rapidly. One system is not necessarily "better" or more effective than another; they meet different kinds of needs, sometimes for different environmental niches or for different types of farmers. Moreover, there are no clear or absolute divides between the formal and local systems: seed and varieties can flow between them; farmers draw upon one or the other depending on need. Seed-related interventions, whether for "relief", "rehabilitation" or "development" need to be based on an appreciation of the strengths and weaknesses of each. As several seed specialists advocate, we need to develop proactive strategies to integrate the strengths of both systems.

SEED SECURITY IN EMERGENCY SITUATIONS

Seed systems comprise both material entities (crops/varieties) and institutional ones (i.e., those interwoven by social networks, markets, companies and seed parastatals). Their complexity and dynamism have been noted - yet so has the possibility of understanding the main characteristics fairly rapidly.

Some of the newly emerging conceptual frameworks that are helping implementers of relief aid to think practically about how to diagnose and assess seed security and farmers' seed needs. The tailoring of the support responses to specific constraints should lead to higher cost-effectiveness of interventions, as well as short- and long-term gains for beneficiaries themselves.

Two broad types of concepts and tools are presented below: a seed security framework (SSF) and distinctions among acutely stressed, repeatedly acutely stressed and chronically stressed seed systems.

Seed Security Framework

Humanitarian agencies working to provide food have long used the food security framework as a tool for thinking about issues related to food security. While the definitions of food security are many and varied (estimated by some to be as many as 200 definitions with some 450 indicators, aspects of availability, access and utilisation (this last aspect particularly associated with quality) often figure as central elements. Formal definitions of seed security are fairly recent, yet these same three basic elements are central to its characterisation. Interestingly, all of these definitions identify, explicitly or implicitly, "access" as the main determinant of seed security.

The first detailed seed security framework, inspired by the USAID Food Security Framework, was developed in 1998 and formally published in 2002. The framework distinguishes among the different types of constraints embodied in the concept of seed security: overall seed availability, farmers' access to seed, and seed utilisation features. These elements are briefly discussed below.

Seed Availability

Availability assessments need to be separated from issues

of the quality of the seed or the desirability of the varieties. If seed that will grow and mature to harvest, but which is of otherwise low quality or of unwanted varieties, is available, this constraint would fall into the utilisation diagnosis. Seed unavailability means just that there is little/no/insufficient seed to plant within a well-defined zone of action. As a basically geographically based parameter, it is independent of the socioeconomic status of farmers.

Seed Access

Access is specific to farmers or farmer groups/communities. It largely depends on the assets of the farmer in question: whether or not the farmer has the cash (financial capital) or social networks (social capital) to access seed. (Arguably, land and physical assets may also be considered as determinants of access: if a farmer has sufficient land to guarantee self-sufficiency and adequate storage infrastructure, he/she is likely to have sufficient access under most conditions.) A diagnosis of a lack of access to seed indicates that traditional methods of obtaining seed have been disrupted and that alternative means are very difficult to employ. Access could be disrupted, for example, if social networks that facilitate seed exchange collapse, local grain markets cease to function adequately, or farmers lose so many assets that they can no longer afford to buy or barter for seed.

Seed Utilisation

The category of utilisation comprises issues of seed quality, including physical, physiological and genetic or varietal characteristics.

In terms of the issues of physical, physiological, and sanitary seed quality, many implementers define quality according to the formal sector definition and thereby equate quality with certified seed. This tendency stems from the requirements of donors and procurement departments to show evidence of formal sector seed certification (and often to furnish a seed grower's certificate) when purchasing seed for aid distribution.

Seed might also be sourced from the local foodgrain markets, then cleaned and sorted. The quality of this seed can be highly variable, and relatively few studies have actually analysed farmer seed quality parameters. Farmers and grain traders may, in some cases, exert considerable effort to distinguish between grain and seed, sorting in the field, in storage and again at sowing time. Farmers' own standards and indicators, however, tend to be different from those of formal sector producers, suggesting that quality assessments may be relative.

Concerning varietal or genetic parameters, formal sector standards tend to aim for varietal purity as being among the more important defining quality characteristics. However, farmers often self-consume the harvest, sell it locally, or sow in a risk-prone and variable environment, which may encourage a certain degree of varietal heterogeneity - rather than purity - as the preferred strategy. So again, assessments of desired quality vary according to the assessor.

It is on the issue of seed quality that the FAO seed relief discussion group had some of its more extensive discussions. There is a fundamental concern among implementers, including FAO, that they are accountable to both donors and beneficiaries for the product they deliver and that, at the minimum, the product has to do no harm. Several practical recommendations on seed quality emerged from the FAO seed relief discussion group analyses: there should be minimal quality standards adhered to - no matter what the intervention context - and it may be possible to introduce simple methods to check these.

The group also suggested a move away from stereotyping: some supposedly certified seed is not of good quality (at least once it reaches the farmer) and some farmer-saved seed definitely is. In addition, good-quality seed can be rendered near useless by poor storage or transport conditions. One basic notion put forward merits further discussion: it is not necessary or appropriate to insist on higher quality than that which farmers would normally be accustomed to in non-emergency situations. The group outlined its recommendations for guiding seed quality parameters in an emergency.

Reflections on the Seed Security Framework

A framework is only useful so long as it stimulates reflection and facilitates appropriate action. One of the accomplishments of the SSF thus far is that it has drawn attention to some of the basic assumptions, conscious or not, on which much of current seed aid implementation is designed. Much of the seed aid response is now based on the premise that seed is not available locally and that the intervention should aim to bring seed physically in from elsewhere to the stressed region. This diagnosis of "lack of availability"(which may be explicit or implicit) has increasingly been questioned.

As with all tools, the SSF is also only as useful as the capabilities of those using it. Parameters of seed security are not static and the lines between them are not always clear. For instance, the distinction between availability and access is dependent on scale. If one is willing to pay enough or to transport seed from far enough away, seed is always available. Likewise, the dividing line between the concepts of availability and utilisation has to be considered in subtle terms. If seed is not really adapted, can we call it seed at all? That is, should seed be considered available if that which is on offer has little possibility of sprouting and yielding, and being acceptable to farmers? The SSF tool has been an important advance in providing analytical concepts by which to start to analyse seed systems, but its use needs to be context-specific.

Acute and Chronic Seed Security Contexts

Much of emergency seed aid to date has been implemented in what are assumed, explicitly or implicitly, to be contexts of acute stress. That is, some outside agent (e.g. a flood) has severely disrupted a livelihood system, including its seed system, and caused a seed intervention to become necessary. Increasingly, however, seed aid is being delivered not as a "one-off" intervention, but on a repetitive basis. In the 1990s, Kenya witnessed at least ten seasons of seed aid distribution, seemingly due to "repeated acute" disasters, mostly drought, with the odd 1998 staccato of the El Niño floods. Similarly, Burundi, Ethiopia, Malawi and Rwanda (among others) have been the scenes of seed aid assistance nearly every season - for five to ten years. Such examples of repetitive seed aid should stimulate reflection and re-examination of the underlying problems.

There are multiple ways to try to characterise the emergency and rehabilitation contexts in which practitioners intervene. Key aspects of disasters have been described through sets of contrasting parameters, which have helped link the context with the planning of reconstructive measures (human-induced/natural, single/complex, rapid/slow onset, short/long duration, heterogeneous/homogenous). Here, for the sake of discussion, we distinguish between two broad categories: "acute" and "chronic." This distinction is not always clear, even conceptually, but it can be a useful tool in informing appropriate seed system interventions.

Acute Seed Insecurity

Acute seed insecurity is brought on by distinct, short-duration events that often affect a broad range of the population. It may be caused by the failure to plant a single season, loss of a harvest, or one-time loss of seed stocks in storage. While in "normal times" one might identify the seed-secure, the semi-secure, and the "always seed-short" households, all may be affected during an acute event, such as a flood or short civil disturbance. Those communities and farmers who recover quickly, with or without one-off seed-distribution assistance, are often those who suffered only from acute stress.

Chronic Seed Insecurity

Chronic seed insecurity is independent of an acute stress or disaster, although it may be exacerbated by it. Chronic seed insecurity may be found among populations that have been marginalised in different ways:

— economically/socially (poverty, little land, little labour)
— ecologically (repeated drought, degraded land)
— politically (insecure areas, on land with uncertain tenure)

Chronically seed-insecure populations may be characterised by (1) continual shortage of adequate seed to plant, (2) difficulties in acquiring off-farm seed because of poverty, or (3) the utilisation of low-quality seed and poorly-adapted varieties on a routine basis. The result is households with a built-in vulnerability to seed system calamities.

Relationship between Acute and Chronic Insecurity

Acute and chronic seed insecurity will very often coexist in an emergency situation. Indeed, in cases where emergencies recur, in drought-prone areas, for example, acute situations are nearly always superimposed on a chronic problem that is rooted in poverty.

An analysis of situations of seed insecurity reveals two trends in the relationship between acute and chronic contexts. First, there is increasing evidence of a transition from acute to chronic seed insecurity rather than recovery, as various forms of "quick relief" (e.g. free distribution of improved varieties) may undermine the function of local seed systems and alter more robust crop profiles. Thus, it may be alarming, but not surprising, that farmers in the Tana Region of Kenya routinely list "seed relief" as one of the basic channels on which they count to access seed, season after season.

Second, analysis is showing that many of the aid cases that were originally considered acute situations, on closer examination exhibit aspects of more chronically stressed populations. Seven of the eight seed intervention cases being examined under a current CIAT/CRS/CN (CARE Norway) project were seen, after assessment, to be primarily chronic situations where acute aid was being implemented. For example, in the Nampula region of Mozambique, provision of clean cuttings of new cassava varieties has helped to address the problem of brown streak virus, which has devastated the local cassava staple crop. However, to strengthen food and seed security in the long term, efforts might need to focus on broad strategies for crop diversification. Fortunately, the local government is moving in this direction.

Certainly, populations are heterogeneous and in every community one might find some seed-insecure households. However, this is different from general community-wide stress. Case studies are showing how, too often, acute emergency measures are being implemented in lieu of possibly more effective and long-term support. This is not a new insight but it is one that has yet to change major emergency seed intervention practices.

Chronic seed insecurity will nearly always manifest itself as a problem of seed access; lack of the ability to access a good is usually equated with problems of basic poverty. One consequence is that the scope of the assessment needs to go well beyond any seed system diagnosis, towards a set of approaches that will support and strengthen basic livelihood strategies and that will help communities become more resilient in the face of real system adversity.

Selection of Appropriate Interventions

If seed availability is identified as the problem (whether acute or chronic), seed-based interventions, such as seed importation or development of community-based seed production enterprises, may be appropriate. However, a diagnosis of "seed access" might wisely trigger a more holistic analysis of livelihood strategies.

In the acute phase, interventions based on seed systems may still largely be appropriate for an access problem, particularly those that can also stimulate the local economy. However, identification of chronic access problems should lead practitioners to reflect on the need for approaches that address livelihood security more broadly. Either individually, as a subpopulation, or as a community, beneficiaries may require broad-based poverty-alleviation assistance. Again, seed-based interventions may help but, in themselves, are unlikely to be sufficient to put stressed populations on a real road to recovery.

Seed Systems in Stress

Rigorous studies of the seed systems that farmers use are relatively new, with most studies carried out in the last ten years. Studies focused on seed systems in emergency contexts are still fewer in number, being limited to a handful of cases. In depth accounts by practitioners can also provide valuable lessons, but these are also the exception rather than the rule.

This relative lack of analysis is disturbing given the level of seed aid. Emergency programmes normally do not build in a diagnostic or assessment component and at the other end of the project cycle, evaluations beyond the harvest are rare. This gap means that interventions aimed at rehabilitation usually cannot profit from what has been recently learned on the ground. The problem is compounded as institutional memory tends to be divided between those involved in "relief" and those involved in "rehabilitation"and "development".

Despite their paucity, a number of findings are emerging from this body of evidence:

— Farmers who have the conditions to sow (e.g. who are not displaced or landless) generally find ways to obtain seed of at least some key crops during crises. In both acute and more chronic contexts, the remarkable resilience of the local seed system has been documented. For example, in Rwanda in 1994, just three months after widespread killing and displacement, with violent events peaking near harvest time, farmers managed to plant their staple crop, beans, in areas comparable to those during the pre-war period. They sourced seed from a range of channels, including their own stocks and local markets. Similarly, data from Sierra Leone, documenting a period of four years of civil strife, showed farmers in a war zone actually increasing their production of grain staples.

— To date, in contexts of emergency seed delivery, where total seed sown has been monitored, analyses have shown that aid does contribute to seed needs, but not in a major way. In Kenya during the 1997 drought, despite massive distributions of maize seed, over 85 percent of the seed sown came from local channels. There were similar findings with regard to bean seed in Honduras after Hurricane Mitch in 1999 and sorghum in drought-stricken Somalia. This does not mean that farmers do not value or need the emergency distribution; rather, it demonstrates that they also value highly the seed they can source themselves - of known varieties and from known channels. Emergency aid adds support to a seed system that may be weak but that still has much to offer. The major exception to this finding concerns situations where the seed on offer is of a crop that is relatively new to the farmers.

— Local grain markets are of central importance as sources of seed. This is one of the real surprises arising from the studies of emergency seed systems. As home stocks of self-saved seed go down, it is the local markets that primarily fill in the gaps, at least in areas where local markets have had some prior importance. From social exchange networks, farmers may receive a handful of seed to show "goodwill," but rarely the quantities needed to sow significant areas.

— Different seed channels may be affected by the crisis to different extents and in different ways. Some are more durable than others. One salient example comes from the Rwanda war and genocide crisis. Farmers, particularly in the north, had come to rely on formal sources for potato seed, as well as for accessing new varieties. The war disrupted this supply as early as 1991-92, as the parastatal responsible for bulking up was located in one of the war's first fronts. Development projects, also standard sources for certified seed and improved varieties, phased out activities early as insecurity grew. Local markets, in contrast, the main sources for bean seed, continued to diffuse local grain used as seed(and sometimes emergency seed as grain) during some of the worst of the events. So while potato production virtually collapsed (relying on the formal sector for improved varieties and clean seed), bean seed channels - and production - that were based on local farmers' systems continued largely on course.

It is important to analyse each crisis on a case-by-case basis: the agro-ecological and political context, the crop exigencies and the principles on which the seed system normally functions all vary from situation to situation. In many war contexts, social relationships may be so severely strained that the routine networks of barter and exchange through which seed is distributed are irreparably damaged. The local seed system in Rwanda did not particularly suffer from this problem as it was never firmly grounded on an ethic of gift giving.

The information to date does suggest that one widespread assumption, that of seed not being available in a crisis, frequently does not hold up to closer scrutiny. Seed can be found, if one has the means. Seed access, in contrast, does tend to be a problem again and again. Victims of a disaster may have lost considerable assets (houses, doors, windows, livestock, seed stocks and social assets, such as support networks) and may not have the margin to re-access seeds (from markets) or may have priorities other than buying seed (e.g. an urgent need to hire labour or pay for urgent healthcare).

In brief, the current body of seed system analyses leads us to avoid assuming that "disaster" equates with "need for seed aid". Some sort of seed-system support may be warranted, but the response needs to be more clearly matched with the constraints and opportunities at hand.

Use of Major Options in Emergencies

Possibly because of their roots in food aid, the two major approaches currently being used to support seed systems in an emergency (acute stress) involve getting seed to farmers; i.e. they involve asset transfers. Seed distribution, often coupled with tool distribution ("seed and tools") and the seed voucher and fair (SVF) approaches are described below. Additional strategies for supporting seed systems in stress are then briefly introduced.

Direct Seed Distribution

Direct distribution of seed seems to have been the earliest approach to seed-based humanitarian relief. While it dates back some 80 years domestically and perhaps 20 internationally, direct seed distribution has only come into maturity as a routine emergency response in the last 10-15 years.

Seed distribution is based on the assumption that, because of a disaster, farming communities do not have enough seed. As seed is often brought in from outside the region, a less explicit assumption, guiding procurement in these cases, is that sufficient seed is not available locally. In its classic form, "seed aid" involves the importation of seed; usually from certified seed sources that have become suppliers because of the requirements of donor and procurement departments.

Because of this, the content of seed aid distribution has generally been limited to a narrow range of crops and varieties - those produced by government or commercial sectors and, hence, widely adapted to larger areas. For example, we find much of the relief in southern Africa to be maize based, often hybrid-maize based. Vegetable seed, imported from outside Africa, has also proven to be a prime input into classic seed relief operations. Certified seed of cabbage, kales, onions, etc., are assessed as being useful for stimulating local agro-enterprise, as well as improving local diets.

A number of guidelines have been written to inform better practice for seed provision during emergencies. Some of these aim to be holistic, discussing the rationale for, and varied approaches to, seed aid in the short and longer term.

Some restrict themselves to the day-to-day practicalities of getting seeds to beneficiaries. Still others are tailored to a specific crop or disaster-prone area.

Since then, extensive experience in seed aid implementation has allowed refinements to be made in the approach, enhancing both its short- and long-term effectiveness. Among other changes, there have been moves towards procuring locally appropriate crops and varieties (steered by specific agro-ecosystem assessments) and towards greater flexibility in seed sourcing. Seed companies are sometimes asked to recondition food grains of crops not easily available through commercial channels (such as cowpeas) and local markets might provide some emergency seed provisions, after it has passed some of the standard seed-quality tests.

Evaluating the process of planning and implementing seed aid itself has also become more common and seems to be leading to more efficient operations. While the two major (and routine) complaints voiced by those supporting the approach persist - that crop and variety choice are not sufficiently wide and/or adapted and that seed often arrives late at the farm level - humanitarian practice here is clearly becoming more refined through time.

Ironically, despite a long and widespread history, there are relatively few real evaluations of the effects of seed aid and almost none have been truly independent. Little is known about the impact of the relief seed in helping farmers truly get back on their feet and in promoting system stability. Nor have the longer-term effects of repeated delivery been sufficiently analysed. Recently, however, an edited collection of papers questioning the effectiveness of conventional approaches and proposing alternatives has been published.

Additionally, the international agricultural research centres have developed a consensus view that some seed relief, although well intentioned, may actually undermine the operation of local seed markets, stifle the potential development of commercial supplies and create marked dependencies among the beneficiary population. Recognizing the major shortage of rigorous evaluations, these issues are now in the process of serious examination.

Seed Fairs and Vouchers

The approach using seed vouchers and fairs in an emergency context is a relatively new aid option, dating from about 2000. Neither seed fairs nor vouchers, however, are totally novel, and the seed fair antecedent, in particular, bears some mention.

Seed fairs within agricultural development have a history of being associated with concerns for biodiversity and cultural heritage. Particularly in the Andean region and throughout South and Southeast Asia seed fairs have been promoted to help farmers maintain their seed diversity, raise awareness of its value and encourage the maintenance of associated knowledge. More recently, seed fairs have been held in Africa (Mali and Zimbabwe) mainly to promote on-farm conservation of plant genetic resources, although in limited instances they have also been used to promote new varieties.

Such market-oriented approaches are now being more widely applied in Mozambique and Zimbabwe as well. Within such biodiversity-focused fairs, many varieties are generally exchanged, but with small quantities of seed passing hands. The aim is not provision of seed, per se, but rather transfer of genetic material, with key associated knowledge.

Vouchers have long been a standard practice within the realm of food aid - and beyond. The first suggestion for their use in seed aid assistance emerged in the mid1990s. As shown above, ground-truthing research after the Rwandan civil war found that seed had generally been available within local seed systems, but farmers had few resources to access it. Selective distribution of seed vouchers was recommended as a way for impoverished farmers to obtain seed through these local channels.

This voucher recommendation, together with the assessment of the access problem, led one NGO (CRS) to make a conceptual leap and test voucher use in an emergency context, first in northern Uganda in 2000. The voucher methodology was subsequently combined with fairs to bring potential seed sellers together at the same place in order to simplify logistics and minimize possible misuse of vouchers. Seed vouchers, coupled with fairs, were first implemented in Kenya in 2000 by CRS through an FAO-supported project with financial assistance from the Swedish International Development Cooperation Agency (SIDA).

The premise of the seed voucher approach is that seed is generally available in a region after a crisis (although this assumption should be continually verified through field assessment). Farmers' constraints hinge on their not being able to access the seed, usually because of sharply depleted finances, sometimes combined with the collapse of social networks. The notion of seed fairs draws on the premise that farmers are willing and able to select for themselves the crops and varieties they require for sowing, especially in difficult periods. Seed vouchers and fairs are generally organised for a period of one to two days and are held in multiple local locations.

To date, seed vouchers and fairs have been organised in nine countries (Burundi, Eritrea, Ethiopia, Kenya, Mozambique, Sierra Leone, Sudan, Uganda and Zimbabwe) and are being implemented or supported by a range of agencies (CRS, FAO, ICRISAT, SIDA, the United Kingdom Department for International Development (DFID) and the newest, World Vision). They have been tested for use in both natural (flood and drought) and human-induced disasters (conflicts).

Manuals have recently been issued to describe the overall philosophy behind fairs and to provide specific implementation procedures. Experience has also refined operational logistics: for instance, more vouchers are now issued in smaller units to allow farmers to "purchase" a greater number of crops and varieties (in small quantities) as well as to give them greater leverage for bargaining with sellers. Furthermore, the obligation of those sponsoring seed vouchers and fairs to conduct an exit interview has proved key for building in a quick and reiterative learning process for fair improvement.

Those involved or interested in the seed voucher and fair approach underscore its range of advantages. It not only provides farmers with seed, but seed for the crops and varieties they desire (whether local, improved, or a basket of both). It fundamentally supports two types of beneficiaries: the buyers or seed recipients themselves, as well as the many small traders (up to 500 at an event) who may put on offer some of their own seed, together with that of the family business. The approach seems broadly empowering (allowing even poor housewives to sell), while at the same time giving economic support to local seed system entrepreneurs and private seed company interests willing to compete on a level playing field.

It is particularly the benefits to the regional economy that are now spurring implementers to broaden the use of fairs in chronic stress contexts. Seed fairs will soon be designed specifically to stimulate local agro-enterprise and to bring entrepreneurs (seed and otherwise) together for an "innovation" event. Given flexibility, seed vouchers and fairs could help promote a seamless progression from a relief to a rehabilitation to a development perspective.

The seed voucher and fair approach is not without its critics and the approach does raise a number of basic concerns, two of which are consistently voiced. First is the quality of the seed. Given that sellers from broad venues can offer their wares, how can agencies guarantee that even minimal seed-quality standards are met and that the beneficiary is not being cheated in the process? The second basic issue revolves around logistics and scale. Seed vouchers and fairs are basically a decentralised form of relief implementation. As they have to be organised locally, the scale of impact is circumscribed, e.g. 30 000 families for any one event. Scaling up implies that multiple fairs would have to be organised in order to have an impact on a range of agro-ecological zones and to reach a sizable number of communities.

Direct Seed Distribution and Seed Vouchers and Fairs

Both approaches, seed distribution and seed vouchers and fairs, are presently in use. Even now, both are being implemented by the same donors/agencies. Both are sometimes programmed within a single country. Initial evaluations comparing and contrasting both approaches have started. Without determining that one approach is "better" than the other, we might consider the appropriateness of each in different contexts and conditions.

REFERENCES

FAO, "Restoring farmers' seed systems in disaster situations. Proceedings of the International Workshop on Developing Institutional Agreements and Capacity to Assist Farmers in Disaster Situations", *FAO Plant Production and Protection Paper*, No. 150. Rome, 1999.

Preston, S.R., "Checklist for use by potential donors before giving seeds to Pacific island countries after emergencies", *Phase II. PRAP Project 03. PRAP Leaflet No. 21*, Suva, Pacific Regional Agricultural Programme, 1999.

16

Seed Processing and Storage

Seeds are fragile, living organisms, and the shelf life of the seed is affected at the beginning of the plant life cycle by such factors as soil nutrition. For example, if the soil is zinc deficient, the quality of the seed will be adversely affected. Though providing the best conditions for crop growth and health is the foundation of seed quality, the factors that can have the most important effect on seed viability and vigor are harvesting, extraction, cleaning, transportation, and storage. It is easy for seed to become damaged at any of these stages.

SEED HARVESTING AND EXTRACTION

Seed harvesting and cleaning methods can be divided into two methods: dry processing and wet processing. Dry processing involves harvesting seed that has already matured and dried within the seedbearing portion of the plant. Examples of dry processed seed plants include beans, broccoli, corn, lettuce, okra, onions, sunflower, and turnips. Wet processing is used when the mature seed is enclosed within a fleshy fruit or berry. Examples of wet processed seed plants include cucumbers, melons, and tomatoes. Some vegetables can be either dry processed or wet processed, for example, peppers, and squash.

Harvesting

The basic rule c. harvesting is to allow the seed to mature as long as possible on the plant without the seed or fruit becoming diseased, or overly ripe. Each type of plant has an optimum time for collecting the seed, but factors such as climate, weather, disease, insects, birds, or predatory mammals may require that the seed be collected at less than the optimum time. In the Mid-Atlantic and South, frequent and daily thunderstorms and high humidity may play a large role in determining how and when seed is harvested. For example, in dry climates, beans can normally be left to mature and dry in the field, but during wet humid weather, it is best to harvest early and allow the beans to continue maturing and drying under cover.

Dry Seed Processing

When seeds are ready to be processed, the entire seedpod, capsule, or seed head will become brown and dry. During the maturation process, the ripening pods and capsules change color from green, to yellowgreen, to yellow, to light brown, to a darker brown, or dark gray. Ripening and maturation may be uneven within the pod or capsule, uneven on the plant, and uneven within the stand of plants. For that reason, the pods of many plants are harvested individually. Seeds of legumes and brassicas often develop a split along one side of the pod. This is the best time to collect the seed, before the pods start to open and scatter their seed. Most flower seed heads are not ready to harvest until the flower head has dried completely to the base, including a short section (approximately ¼") of the supporting stem.

Some plant families, such as the Asteraceae (Aster family) have a smaller percentage of viable seed in the head, and the seeds continue to mature after collection. For this reason it is best not to be too hasty in harvesting the seed.

Examples include lettuce and sunflower. Some seed may mature in the capsule or pod, even before the pod has turned completely brown. Most seeds turn a darker color as they mature. Seeds may initially be white, turning green or tan, and then brown or black. Once the seed pods, capsules, and seed heads start to mature, it is important to check the crop on a daily basis. Rain or seed predators can ruin a good seed crop in a short period of time.

Plants that produce umbels (members of the carrot family, or Umbelliferae) can usually be left in the field to harvest until the umbels are dry. Some members of this family mature their seed unevenly causing seed to scatter, while other seeds in the umbel continue to mature. One method of dealing with crops that mature their seed unevenly is to pull the plants and hang them upside down to dry under cover. This allows the seed to continue to mature on the plant while the plant dries. This procedure is often used for lettuce. Confidence in knowing when to harvest comes both with experience and familiarity with different species and crops.

After harvest, seeds are threshed to remove the seed from the surrounding plant material. A period of air-drying is important before seeds are threshed. Plant material should be spread out in thin layers until all plant material is dry; otherwise, mold, decay, and heat from decay will cause damage to the seeds. As the plant material dries, seed pods may split open or shed seed. Harvested material should be stored in a wellventilated room with low humidity. During this time you should be aware of insects, especially weevils that feed on the seeds. Plant material that is ready to be threshed should be brittle.

Threshing is best done outside on a dry day. The threshing process involves application of mechanical force using a controlled pressure and a shearing motion., and is accomplished by hand or by machine. There are many different methods for threshing seed. Plants that have pods, such as beans and okra, can be threshed by placing the pods in a large feed sack, which is tied shut securely, and then placed on the ground where it is flailed, stepped on, jogged on, or danced on with a twisting motion. The sack is turned often to redistribute the plant material for further threshing. When using this method, it is best to use running shoes or other soft-soled shoes because seeds can develop hairline cracks and splits from too much pressure.

To separate seed from flower heads, plant material is spread on a concrete slab and then gently walked on with a twisting motion to break open the flower heads. Care must be taken to not apply so much pressure that the seed is abraded or broken (especially a concern with angular-shaped seed). To thresh seeds of brassicas, place plant material in a large wheelbarrow or on a large tarp, and while wearing gloves twist and wring the plant material through your hands until the seed breaks free from the pods. Another method for extracting seed is to place seed heads on a piece of plywood.

The seed is extracted by placing a wooden cement float above the material and then pressing and twisting until the seed breaks free. If the seed heads are small you can run a rolling pin over them. For threshing small lots of seed, a threshing box may be used. This consists of a wooden box, with sides slanting outward, open at the top and on one end, and the long sides and bottom covered with corrugated-rubber floor matting. Seed is placed in the box and a rubbing board (also covered with corrugated rubber), is moved back and forth across the seed heads to separate the seed from the heads.

Wet Seed Processing

Wet seed processing is used with seed crops that have seeds in fleshy fruits or berries. There are three steps to the process: (1) extraction of the seed from the fruit, (2) washing the seeds, and (3), drying.

Extracting seed

The type of extraction process depends on the species. Soft fruits such as tomatoes are cut up, mashed, and then fermented. Cucumbers and melons are cut in half, the seed scraped out along with the fruit pulp surrounding the seed, and then fermented. In watermelons, the entire fleshy fruit is fermented along with the extracted seed. These types of fruits have a gel surrounding the seed that contains germination inhibitors. The presence of the gel also makes handling and drying of the seed difficult. Fermentation is a natural process that occurs to a small extent as fruits decompose. When fermentation is done in a controlled manner, the microorganisms, principally yeast, break down the gel thus releasing the seed while killing bacteria and fungi that cause most seed-borne diseases. The temperature and length of fermentation are important. If the mash is not fermented long enough, seedborne diseases will not be eliminated, but if fermented too long, the seeds may sprout prematurely.

The length of the fermentation is dependent on temperature and typically last three days at a temperature of 70 to 75°F (21°C to 24°C). Length of fermentation may also depend on the variety itself. For example, varieties with high sugar content may take longer to ferment, up to four days. With few exceptions, fermentation periods longer than three days risk damaging the seed. There are different fermentation techniques for different crops, for example, pepper seeds are extracted from the fruits by mashing, but the fermentation process may last only 24 to 48 hours. Though eggplant isn't a watery fruit, it can be mashed and fermented for about 48 hours.

A Corona® grain mill is a very handy tool to use for extracting seed of eggplants and hot peppers. The grinding plates on the mill are set wide enough apart to break up the fruit without grinding the seed. Small peppers can be put directly into the mill, whereas larger fruits such as eggplant are cut into small cubes before processing. Before processing another variety, the mill has to be disassembled and cleaned thoroughly to remove bits of small seed caught in the plates and other parts of the mill. Though some people use a food processor and blender set on low speed to extract the seed of hot peppers and eggplant, it is easy to damage the seed.

Instead, a food processor that has a thick (approximately 1/8") blade works well provided there is ample water in the processor, the speed is as slow as possible, and the processing time as short as possible. There are some differences of opinion about adding water to the fermenting mixture, the concern being that water slows the fermentation process causing premature sprouting. In my experience this hasn't been a problem. As long as the ferment is not diluted more than perhaps 10 to 20 percent by volume, this isn't a concern. The issue of whether to add water depends on how thick the mash is and it depends on the variety.

Some varieties make a very thick mash that is hard to stir, and others make a watery mash that stirs easily. The mash should be stirred three times a day, once in the morning, once in mid-day, and once in the evening. If the mash is too thick to stir easily, the nutrients are not going to circulate easily. Stirring is also important for better control of seed-borne diseases. In addition, when the mash is not stirred, a foul-smelling, white mold forms on top of the mash. This mold can discolor (darken) and damage seeds at the top of the mash. These seeds will later have to be hand picked out of the dried seed. A properly fermenting mash should not have a foul smell, and there should be little or no white mold on the top. A small amount of white mold is not harmful and can be stirred back into the mash, but a heavy overgrowth should be removed.

Washing seed

After fermentation is complete, the seeds are washed to remove pulp, pieces of fruit and debris, and low quality seed. Before washing the seed, it is useful (especially for washing tomato seed) to first scoop out pieces of pulp floating on top of the mash. This is done by straining the mash with your fingers, pulling out the larger chunks. Whether or not there is floating pulp depends on the variety or how thoroughly the fruit was processed. Add a volume of water at least equal twice the volume of mash. It is important to dilute the mash sufficiently because the more dissolved solids there are in the mash, the higher the specific gravity. If the specific gravity is high (lots of soluble solids) it will be more difficult to wash the seeds properly. As a general rule, good seeds are heavy and sink to the bottom, whereas poor quality seeds are light and tend to float off with the wash. The washing process is repeated until the wash water becomes clear.

Although most good seeds sink to the bottom, some vegetables have very light seed and require extra care during the washing process. For example, it is common for a significant amount of good pepper seed to float rather than sink during the washing. This can be avoided by adding the wash water slowly, so as not to create tiny air bubbles that adhere to the seeds, making them buoyant. Even with this precaution, there are a few varieties of watermelon for example, where the good seeds tend to float rather than sink, thereby requiring special care in washing. In this case, the best way to wash the seed is to pour the wash through a ¼" hardware cloth screen, and then use a hose to force pulp through the screen.

Drying seed

Seeds should be dried fairly quickly after washing. Slow drying may result in mold growth or premature sprouting of the seed. In the Mid-Atlantic and South, seeds should not be dried in the sun, nor should they be dried anywhere where the temperature exceeds 95°F (35°C). Dark colored seeds are especially vulnerable to damage when sun dried. Instead, seeds should be dried in a climate-controlled environment using fan ventilation. A combination of ceiling fans and air conditioning dries seed safely and very quickly. Seed should be spread out in thin layers (no thicker than ¼" for small seeds) and then stirred several times a day until dry. Once the seeds feel dry, they should cure for another two to three weeks.

Curing is the final stage in the drying process. As the seed moisture content declines it comes into equilibrium with the relative humidity. After the seeds are cured they can be placed in a container. When drying seeds, choose plywood, window screen, or any hard, non-stick surface. Avoid using paper towels, newspaper, cardboard, or cloth because seeds will tend to stick to the surface making them difficult to remove. Beginning growers often make the mistake of drying squash seed on newspaper, which adheres permanently to the seed coat. No one wants to read the daily news on the surface of squash seed.

SEED PROCESSING AND HANDLING

Threshing Equipment

For small-scale seed production, it is often not cost effective to purchase threshing equipment because of the small volume of seed produced. A number of small alternative seed companies, specializing in heirlooms and specialty seed, thresh most, if not all of their seed by hand. Some seed crops, especially herb and flowers seeds are produced in such small quantity (and have such small seeds) that it is impractical to use a machine for threshing. Also flowers and other seed crops, if left to dry on canvas or a tarp, naturally dehisce (shed) much of their seed. The remaining amount of seed can be released by flailing or rolling, or other non-mechanical means.

It is difficult to locate low-cost, low-tech shredders for small-scale seed threshing. The options are to locate old, used equipment or to construct your own. Vegetable seeds, such as beans can be threshed with a modified leaf shredder/chipper. Typical threshing rates for this machine are 22 to 30 pounds/hour for brassicas, and approximately 100 pounds/hour for beans.

Seed Cleaning Equipment

It is not necessary to have expensive seed cleaning equipment to clean seed for small-scale production. With the exception of lettuce seed and a few others , the majority of crop seeds can be cleaned with homemade seed screens. Winnowing will still be necessary to remove smaller chaff. Many seeds can be screened with several different mesh sizes of hardware cloth. Hardware cloth is readily available in the following mesh sizes: ½", 3/8", ¼", and 1/8". The ½" and ¼" mesh sizes are available at most hardware stores. The 3/8" and 1/8" mesh sizes are the most useful sizes, but are the most difficult to find.

Aside from hardware cloth, a lot of other materials are useful for making seed screens. Aluminum window screen can be used for small seeds. You'll also find various meshes of screen available in the housewares section of department stores. Cabinets from electronic devices often have round or oblong holes which are useful for cleaning some types of seed. Special meshes can be ordered from mail-order hardware specialty catalogs. It is useful to collect a large variety of mesh sizes and shapes to handle a wide variety of seed types.

Once you have located suitable screen material it should be mounted on a frame. Professional seed screens are mounted on 12" square wooden frames. Homemade screens should also be mounted on 12" square frames so they can be used together with professional screens. The frames can be made of wooden lath which measures ¾" x 1-1/2". Ideally, the screens should be constructed so that they nest together. The nesting feature is desirable for using three screens simultaneously: rough chaff is retained on the top screen, seeds in the middle screen, and small chaff and small seeds on the lower screen. Once the frame has been made, the screen is nailed with small nails or brads at approximately 1" intervals all along the edge of the frame.

Winnowing Equipment

The classic method of winnowing involves placing seeds in a wide basket and tossing the seeds and chaff into the air. The chaff is carried away by the wind. Illustrations of Native Americans using this historic method evoke a bucolic mood, but in reality, winnowing by this method is extremely difficult, and the results are not very satisfactory. The most vexing part of the process is that wind speed is always changing in velocity and direction. It can work for certain kinds of seed, but it actually works better to use two large bowls, pouring seed from one bowl into another bowl below, while blowing on the plant material as it falls. This method works satisfactorily if the seed is heavy and the chaff is very fine, and susceptible to being carried away by a gentle current of air. In any case, hand winnowing should be done, not on windy days, but when the air is calm.

A good range of equipment for winnowing includes the following: (1) an assortment of stainless steel bowls ranging in size from 6 to 16" in diameter, preferably bowls with varying shallow and deep sides; (2) an electric hair dryer with two speeds (heating element removed, or unheated air setting); (3) portable vacuum cleaner with option of connecting hose to the air discharge opening: (4) household box fan with three speeds; (5) squirrel cage blower; (6) rheostat for controlling fan motor speed; (7) a tarp for catching seed or chaff; and (8) a dust mask for keeping chaff out of your lungs.

Most light seed can be winnowed quickly and efficiently with mixing bowls and a hair dryer. It takes a little practice determining the correct amount of seed to put in the bowl, setting the proper fan speed, and the distance of the dryer from the bowl. It helps to jiggle the bowl up and down or to swirl the seed in the bowl. There must be little or no wind when you work with light seed. Until you have a little practice it is best to put a tarp on the ground, in case you need to sweep up your mistake and start over.

Heavy seed is best winnowed by pouring the seed from one container to another in front of box fan, or alternatively directing the air discharge hose from a vacuum cleaner into a large (16" diameter) mixing bowl. When dealing with large volumes or certain types of seed it is helpful to use mechanical equipment for winnowing.

SEED DRYING

Principles of Drying:

Drying is a normal part of the seed maturation process. Some seeds must dry down to minimum moisture content before they can germinate. Low seed moisture content is a pre-requisite for long-term storage, and is the most important factor affecting longevity. Seeds lose viability and vigor during processing and storage mainly because of high seed moisture content (seed moisture greater than 18%).

High seed moisture causes a number of problems:

— Moisture increases the respiration rate of seeds, which in turn raises seed temperature. For example, in large-scale commercial seed storage, respiring seeds may generate enough heat to kill the seeds quickly, or to even start a fire if not dried sufficiently. Small-scale growers are not likely to have such an extreme condition, but seed longevity will, nevertheless be affected.

— Mold growth will be encouraged by moisture, damaging the seeds either slowly or quickly, depending on the moisture content of the seeds. Some molds that don't grow well at room temperature may grow well at low temperatures causing damage to refrigerated seeds. In such a case there may be no visual sign of damage.

— Unless seed moisture is at least eight percent or below, insects such as weevils can breed causing rapid destruction of seeds in a short period of time.

Long-term Storage

Silica gel is the most effective desiccant (moisture absorbing material) for drying seeds. Powdered milk has been recommended as a desiccant in older seed-saving literature, but is less than ten percent as effective as silica gel. Silica gel is a highly porous form of silica that absorbs moisture. It is available as a powder or as beads in different sizes. The best size bead for drying seeds is approximately 1/16" to 1/8" in diameter "Color-indicating" silica gel is a form of silica gel that has been treated with a small amount of cobalt chloride which acts as a moisture indicator. When the indicator gel is completely dry it is a deep blue color.

As the gel absorbs moisture from the air, it gradually changes in color from deep blue to light pink. Though silica gel is clearly allowable under the Organic Rules, color-indicating silica gel is not clearly covered. Check with your certifier before using color-indicating silica gel. Silica gel can be repeatedly reactivated (re-dried) after it has absorbed moisture. The procedure involves heating the gel and driving off the moisture, and as it dries the color gradually changes to deep blue. Drying must be done in a controlled fashion, otherwise the beads will turn black, and the moisture-absorbing capacity of the beads will be destroyed. There are two methods for reactivating silica gel:

Oven-drying method

This method gives the best results, but it takes longer and uses more energy. Set the oven for a temperature of at least 200°F, but no higher than 275°F. Remember, that some ovens may run hotter than the set temperature. Place the silica gel in a thick-walled Pyrex dish, no deeper than one inch, and continue heating until the beads turn deep blue. When drying large quantities (a pound or more), the gel should be stirred occasionally. The oven drying method takes 1-1/2 hours per quart of gel (with the oven temperature set at 275°F). One quart of silica gel weighs approximately 30 ounces (1.9 pounds).

Microwave-drying method:

The microwave method works much faster but must be monitored more closely to avoid overheating the gel. Use only a thick-walled Pyrex container for heating the gel. Set the microwave on medium or medium high and dry for approximately three to five minutes. The color change in the gel can be monitored through the over door, but the gel should be inspected and stirred at the end of each heating cycle. If the gel has not dried, heat again for another three to five minutes. Approximate drying time is eight to twelve minutes per pound of gel, though actual heating time will depend according to the type of microwave.

There are some fine points and cautionary notes about using silica gel. When using a Pyrex dish, the glass should be thick. Do not use plastic microwave containers that will melt on contact with the gel. Silica gel gets very hot, and the glass container may shatter if it is too thin, of the wrong type, or if unevenly heated. You may notice a slight odor during heating. This is due to either overheating the silica gel, or to the evaporation of organic seed volatiles absorbed by the silica gel during the drying process. Silica gel itself is chemically inert, non-toxic, non-corrosive, and odorless, but breathing the dust can be hazardous under prolonged and repeated exposure. For that reason, do not use the finely ground or powdered form.

Instructions for Drying Seed with Silica Gel

— To dry seed, determine the weight of the seed to be dried, including the packets or envelopes that contain the seed. Measure out an equal weight of silica gel and place the seeds and silica gel in an airtight container for seven days. When drying seed it is important to keep the container size small in relation to the volume of seeds being dried.

— At the end of seven days, remove the packets of seed from the drying container and transfer into another airtight container, such as a Mason jar, Seed Saver Vial™ or barrier pouch. Because seeds can re-absorb moisture from the air quickly, they should be transferred quickly. When used as directed, silica gel dries seed from 12% typical moisture content to a desired moisture content of approximately 5% for small seeds and 7% for large seeds. As a "rule of thumb", seeds will not be damaged provided the drying time doesn't exceed seven days. This drying time applies to humid climates, such as the Mid-Atlantic. Longer drying may drop the moisture content below 3% for small seeds and below 5% for large seeds, levels which may damage seeds or force them into dormancy. Legumes, such as beans, are especially are injured by over drying.

SEED STORAGE

Effect of Temperature on Seed Longevity:

The general effect of temperature on longevity is that

longevity increases as temperature decreases. This is true of "orthodox" seeds: that is, most seeds that follow some general "rules of thumb" regarding longevity during the storage life of the seeds. The relationship between temperature and seed longevity is that for each 10°F (5.6°C) decrease in temperature, longevity doubles This rule applies to seeds stored between temperatures of 32°F (0°C) and 122°F (50°C). This rule assumes that the moisture content is a constant. This is a general guideline; in reality the longevity of some vegetable species declines more rapidly than suggested by the rule, while the longevity of others declines more slowly in relation to storage temperature.

The longevity of seeds is generally not affected by subfreezing temperatures provided the moisture content is less than 14% (because ice crystals do not form). This has been established by a number of published reports as well as my own experience storing and germination testing a wide variety of vegetable, flower, and herb seeds at 20°F (-7°C) and below for periods of at least five years.

This is the ideal way to store seed, especially small seed that doesn't require much freezer space. One caveat: seed cycled in and out of the freezer too many times without redrying may cause degradation of germination

Effect of seed moisture and humidity on seed longevity: Seed moisture has a greater effect than temperature on seed longevity (as noted in the previous section titled "principles of seed drying"). Most seeds also follow some "rules of thumb" regarding moisture and longevity. The general relationship is that for each one percent increase in seed moisture, longevity decreases by half. This rule applies to seed with moisture content between 5 and 13%. Above 13% moisture content, seed storage fungi and increased heating due to respiration cause longevity to decline at a faster rate. Once seed moisture reaches 18 to 20%, increased respiration, and the activity of microorganisms cause rapid deterioration of the seed. At 30% moisture content, most non-dormant seeds germinate. At the low end of the moisture range, seed stored at 4 to 5% moisture content is unaffected by seed storage fungi, but such seeds have a shorter longevity than seed stored at a slightly higher moisture content.

Relative Humidity and Seed Moisture Content

When storing commercially grown seed, it is impractical and too costly to use desiccant to dry the seed for storage, unless the seed is small and expensive. Commercial seed is usually packaged for short or longterm storage under conditions of ambient humidity (unless special equipment is used). Because relative humidity has a significant effect on seed moisture content, it is important to understand the relationship between humidity and seed moisture.

Temperature and Moisture on Seed Longevity

The effects of temperature, moisture, and relative humidity were discussed above as separate factors which affect the longevity of stored seed. In reality, the effects of temperature and relative humidity are highly interdependent in their effect on stored seed. There is a simple method for calculating the combined effects of relative humidity and temperature on seed longevity, which is as follows: the sum of the storage temperature (in degrees F), plus the relative humidity (in percent) should not exceed 100. Since seed moisture is the most important concern, the rule stipulates that no more than half the sum should be contributed by the temperature.

The majority of crop seeds lose viability quickly when the humidity approaches 80% at temperatures of 77°F (25°C) to 86 F (30°C), but when stored at a relative humidity of 50% or less, and a temperature below 41°F (5°C), seeds will remain viable for at least ten years. If seeds are taken from a cold or frozen storage and transferred to room temperate, care must be taken to prevent condensation on the seeds. If the seeds are in a sealed container, allow them to sit until they reach room temperature before opening the container. If they are stored in paper, place the seeds into a plastic bag with the excess air sucked out, seal the bag,, and wait for the temperature to stabilize before unsealing.

Illumination

The effects of light on stored seed have been studied (including the effects of different wavelengths of light): Some studies showed a benefit and some showed a detriment: the results are inconclusive and controversial.

Seeds stored in glass containers should be stored out of direct sunlight because of the localized "greenhouse heating effect" on seeds. This might seem an unnecessary and overly obvious cautionary note, but in my experience, it has happened accidentally more than once. For example, this can happen if jars of stock seed are taken outside for the purpose of removal of some seeds for planting, or for transporting to another location. Though the jar may be temporarily stored in the shade, the angle of sunlight may change quickly and the jar will be in direct sun causing very rapid heating within the jar.

Though some commercially produced seed is dried in direct sunlight (in dry climates), drying seeds in the sun is a questionable practice in the Mid-Atlantic and South if the air temperature is above 90°F (32vC). The air temperature at the seed surface is higher because of the conversion of light energy into heat at the seed surface, and the heat is "moist heat" (though this wouldn't be in an issue in dry climates where evaporative cooling occurs at the surface). I'm not aware of any studies on this issue for our region. Another concern, also not well documented, is that the ultra-violet light from the sun may have a deleterious effect on seed longevity while the seeds are drying. Harrington's suggestion was based on the known effects of ultra-violet radiation on biological systems (rather than specific data). Whether the ultra-violet exposure is long enough to cause concern, is unknown.

Respiration and Heating:

The largest factor affecting respiration and heating is moisture content, and therefore, at a minimum, seeds need to be kept dry. Respiration in seed storage has three effects:

1. Depletion of food reserves. Over the life of stored seed, depletion of food reserves is inconsequential.
2. Release and accumulation of gasses that may affect viability of seeds in storage. The accumulation of respiratory end products such as carbon dioxide is an advantage of storing seeds in sealed containers where carbon dioxide replaces oxygen in the air thereby slowing down respiration and increasing longevity.
3. Release of energy, mostly in the form of heat. When seeds are stored under favorable conditions, respiration is of little consequence. When moisture is high, respiration increases, which in turn increases the production of heat thereby decreasing longevity.

Effects of Fungi, Bacteria, and Pests on Seed Storage

The process of seed harvest and cleaning removes most debris and insects, but certain fungi, bacteria, and insects make their way into stored seed. Fortunately, the same conditions that are favorable to seed preservation inhibit fungi and bacteria and kill insects.

— *Bacteria* - Bacteria do not have a significant role in seed deterioration because free water is required for bacterial growth, and if the moisture content of the seed is high enough to support bacteria, the seed is more likely to succumb to deterioration due to other causes such as fungi, respiration, heating or premature sprouting.

— *Fungi* - Most seed storage fungi are inhibited when the relative humidity is kept below 65%. At this relative humidity the moisture content of starchy seeds is about 13%, and oily seeds about 7%. The major effects of fungi are to:
  — decrease viability; o produce toxins that affect seed viability and germination;
  — increase heat production - important in large seed lots; and,
  — cause discoloration, mustiness, and caking.

— *Insects* - In hot, humid climates such as the Mid-Atlantic and South, mites, weevils, flour beetles, and borers can be a serious problem in stored seed, but if the seed is dried to 8% moisture content and the temperature reduced to (64 to 68°F (18 to 20°C), insects should not be a problem. At a moisture content of 15% and a temperature of 86 to 95°F (30 to 35°C), they can become very destructive. Mites will not survive when the relative humidity is below 60%.

Variety and Harvest Conditions

Different varieties of a particular crop species may have different longevity when stored under the same conditions. This may be explained by differences in enzyme activity, or differences in the chemical constituents of unique varieties. For example, sweet corns are rich in simple sugars, whereas dent corns are high in complex carbohydrates. This explains why dent corn seed keeps longer than seed of sweet corn. Harvest conditions may also affect longevity of seeds in storage. For example, under poor conditions of harvest, seed may have more fungi on the seed coat. Differences in harvest conditions are insignificant if storage conditions are good, but under adverse conditions, harvest differences can be very significant.

Long-term Sealed Storage

Seed can be stored for long-term sealed storage provided that the seed moisture content is less than 8%, which means that the relative humidity must be kept below 35%. For the Mid-Atlantic and South, seed must be dried with a desiccant in order to achieve the desired moisture level for long-term storage. When drying with silica gel according to the procedure described above, large seeds such as peas, beans, and corn will be dried to about 6 to 8% moisture content, and smaller seeds will dry down to about 3 to 5% moisture content, depending on the size of the seed and whether its primary food reserve is starch or oil. Seeds with a starchy food reserve retain a higher seed moisture content than oily seeds.

Packaging Materials and Containers for Seed Storage

Materials for short-term storage

There are a wide variety of materials that can be used to store seed for short-term storage. Most of these are non-rigid materials such as cotton, burlap, paper, and composite materials such as multi-wall paper and plastic film, or polyethylene bags. Materials used for short-term storage are generally porous. They adequately contain and protect the seeds from mixing, but do not provide protection from moisture or loss of seed viability. Such materials are usually used for mechanically separating seed lots, and for transporting and shipping seed until the seed can be placed in environmentally controlled conditions for longer-term storage. Each type of packaging material has its own advantages and disadvantages. Burlap bags have the greatest strength, can be re-used many times, and can be stacked high without slipping of the stack.

The strength of cotton bags is dependent on the thickness of the weave and thread, and the quality of the seams. Though not as strong or as tear-resistant as burlap, cotton bags can often be re-used, depending on the quality of the fabric. Bags made of woven plastic material also are fairly strong but tend to slip when piled high and are harder to close securely after the bag has been opened. Multi-wall bags are made of several layers of paper in a variety of types of construction. They have poor bursting strength when piled high or accidentally dropped, and if used repeatedly, they tend to become brittle along the folds and wear points.

Cardboard boxes and cans, though expensive, are re-usable, good for stacking, and provide some protection against mechanical injury to seed and to infestation by seed storage insects. Flexible packaging that has a weave, whether it is burlap, cotton, or plastic offers little protection against seed storage pests such as grain moths. Consequently, seed stored in such woven bags may have to be inspected if an outbreak is detected in a particular seed bag. Seed storage insects are very good at locating small openings in bags or containers that are not well sealed.

For small lots of seed, paper bags such as lunch bags are inexpensive and adequate for storing seed, but the seams are not always reliable, and when used, the bags should be double or triple bagged to ensure integrity to prevent bursting. Other materials such cellophane, acetate, and 2- to 4-mil polyethylene zip lock bags may be used (provided the seed has dried adequately). If using polyethylene, it is best to use the 4-mil thickness, especially for heavy seeds. Plastics and thin films are not reliable moisture barriers, though such materials offer better moisture protection than paper.

Materials for long-term storage:

Metal and glass containers, properly sealed to prevent the exchange of moisture and gas, are the most commonly used containers. They are the only reliable means of protecting seeds against humidity, insects, rodents, floods, and mechanical damage. Plastic should not be used for long-term storage. For storing large quantities of seed, metal five-gallon cans fitted with a rubber gasketed lid and pressure ring are ideal for storing large seeds such as peas, beans, and corn. One-gallon jars are also excellent, provided that the lid has a gasketed seal. Some one-gallon jars have plastic lids that are flexible enough to form a tight seal when the lid is screwed tightly, but metal lids need a rubber gasket. Gaskets can be cut from sheets of neoprene rubber (available at hardware stores) or used automobile inner tubes. Though glass is breakable, it has the advantage of being transparent so that the contents can be easily inspected for insect damage (especially useful for detecting bean weevils, grain moths, and other seed pests).

Metal and glass storage containers are available from a number of sources. Glass canning jars are available in sizes ranging from ¼ pint to ½ gallon. Onegallon glass jars are often available free for the asking at restaurants, fast-food restaurants, and snack bars that purchase pickles, relish, mayonnaise, and a number of other food ingredients in bulk. For storing small quantities of seed, ¼ and ½ pint jars are ideal, provided the rubber seal is in good condition. Baby food jars are often recommended, but again the seal should be in good condition.

The seal can be ruined from seeds getting trapped between the rim of the jar and the lid; therefore jars should not be packed to the rim. Other materials that can be used are Seed Saver Vials™which are made of a high-density polypropylene with a patented inner and outer valve that ensures an airtight seal. Another material, often used for seed banks, is a heat-sealable barrier pouch. Barrier pouches are a triple-laminate material made from paper, plastic, and foil. See the Supplies and Equipment section for suppliers. They are sealed with a heat sealer, or an iron set on "wool" setting. When deciding on the type of storage container, it is often helpful to settle on a minimum number of standard sizes. If using canning jars, it is best to choose a particular brand name and style so that samesize and same-style containers have the same weight. This is very helpful in inventory control of seeds being held for long-term storage. Standardization of container style and weight makes it easy to obtain the weight of stored seed without actually removing the seed from the container and exposing to humid air. Once the weight of standard containers is known, it is easy to subtract that weight from the container of seed to obtain the weight of the seed.

SEED TREATMENTS

The ideal seed treatment should be: (1) very effective against seed-borne pathogens, (2) relatively nontoxic to animals and plants, even if misused, (3) effective for a long time during seed storage, (4) easy to use, and (6) economical. The hot water treatment method meets many of these criteria. Though not as easy to use as chemical treatments, it can be more effective and is non-toxic.

The purpose of the hot water treatment method is to submerse seed in water hot enough to kill the pathogen without damaging the seed embryo. When properly done the procedure is very effective, but there are some precautions to follow: (1) use only new, vigorous, high germination seed, (2) control the temperature of the process very carefully, since this is fundamental to success, (3) test the germination of the seed lot before and after treating the seeds, and (4) practice first on a small sample before treating a large batch of seeds. Although the germination of properly treated seed will not be significantly affected, the storage life of the seed may be shorter.

Equipment need for this process is as follows: (1) large metal pot and several smaller containers; (2) bags for seed made of muslin, cotton, or nylon (panty hose will do); (3) an accurate thermometer from a science supply house, and

(4) sieve, colander, vegetable steamer, or hardware cloth cut in the same diameter as the inside of the pot.

The procedure is as follows:

1. Place the seeds loosely in a fabric bag, filling no more than half full, and then tie securely. Add a small weight, such as a bolt to help sink the seeds.
2. Set up the treatment container so that the bag of seed doesn't touch the bottom of the pot. There are several ways to keep the bag off the bottom: (a) suspend the seed bag on a vegetable steamer, (b) set the bag in a colander, or (c), cut a circle of hardware cloth and suspend it at least ½" off the bottom by using small stones.
3. Fill the treatment container about half full, and using an accurate thermometer, bring the water to the treatment temperature.
4. Pre-soak the seeds in water, between 100 to 105°F (38 to 41°C) for five to ten minutes to prewarm the seeds.
5. Once the temperature in the treatment container is constant, add the bag of seed. While treating the seed, stir the water slowly and constantly, while keeping an eye on the temperature and the clock. An accurate thermometer is essential because temperatures one or two degrees too high can injure the seed, and one or two degrees too low will fail to kill the pathogen. To help adjust the temperature, cold or hot water can be added while stirring.
6. Remove the seed, and cool by dipping the bag briefly in lukewarm water, and then in cool water.
7. Open the bag and spread out the seed to dry on a hard surface or fine screen. Dry at room temperature using a fan for supplemental ventilation, and stir the seeds several times daily until dry.

GERMINATION TESTING

Most growers prefer to sell seed wholesale and leave germination testing to the seed company that has the equipment and supplies necessary to do accurate testing. You can perform your own germination tests for the purpose of testing the germination of old seed lots, or testing new seed lots that you may want to store for some period of time. Tests that you do on your own may not necessarily be legal for the seed trade, but they will be accurate enough for your own use.

For supplies you will need blotters or thick paper towels, tweezers or forceps for picking up or placing seed, zip lock bags, a misting bottle, and perhaps plastic boxes used for honeycomb (for light-dependent germinators. As a substitute for a germination chamber you can place the seeds on top of a refrigerator or hot water heater. capacity) and remove the compressor. Alternatively, an insulated plywood box can be used as a chamber. Cover any holes with caulk, duct tape, or other suitable material. For monitoring temperature, drill a hole in the side or top to insert the probe of a thermometer.

A maximum-minimum thermometer is recommended because it records the temperature history and is useful for determining if the temperature has exceeded the desired range. For a heat source, use a 40-watt bulb in a porcelain fixture and wire it in series with a computer fan and dimmer switch to control the heating rate. The dimmer shouldn't be run below 75% of capacity because the computer fan may stall or not start up when the current is switched on. The electrical supply cord from the chamber is connected to a timer to give a 16-hour warm period at 86°F (30°C) and an 8-hour cool period (ambient room temperature).

To perform a simple germination test, count out 100 seeds and place them somewhat equidistant from each other, usually not closer than ½", spreading them out on the upper half of the moist blotter or paper towel. The bottom half of the blotter is folded over the top, and then the towel is folded over with a fold about every two to three inches and held in place loosely by a rubber band. The towels or blotters are then placed vertically in plastic trays refrigerator containers, which are placed in the germination chamber or suitable location. The trays or containers should be covered loosely with zip-lock or plastic bags to keep the moisture in the towels, but they should not be completely closed because seeds need air. Towels are inspected twice a day and misted as necessary.

Most vegetable seeds germinate in about 7 to 14 days. During the first count remove the seeds that have germinated and when the final count is done, the remaining seeds are scored and the paper towel discarded. Seeds that require light are placed on blotter paper within plastic honey boxes. The light from the germination chamber will be sufficient to satisfy the light requirement. Some cool-weather crops such as lettuce should be germinated at room temperature. Most varieties of lettuce will not germinate above 80°F (27°C). Although a test of 100 seeds will give you a fairly reliable germination test, a test of 400 seeds will remove most of the statistical error and sampling error in testing.

Germination Enhancement Techniques

There are several techniques that can be used to improve the germination of seed lots. These are especially useful when it is necessary to bring a lot of seed above the Federal germination standard.

Mixing seed lots of the same variety

The easiest way to enhance germination is to combine two lots of germination tested seed, one having a high germination score and the other having a lower germination score. The two lots are thoroughly mixed and the new germination percentage is calculated. For example, mixing 600 grams of 90% germination seed with 400 grams of 70% germination seed yields 1000 grams of 82% germination seed. This technique might be used by a seed company in an effort to meet market demand when the seed is in short supply or to avoid discarding seed. If this technique is used, the seed lot with the lower germination should not be too low in vigor or germination, otherwise there will be too much variability in the stand of plants produced from such seed.

Additional seed processing

Types of seed that have been cleaned using winnowing as part of the seed cleaning process are candidates for these methods. Seeds of pumpkins, squash, melons, cucumbers, sunflowers, lettuce, and flower seed can be aggressively re-winnowed to remove lighter, immature seed that is not well filled out. Such seed will be lighter than the good seed, which is denser and more resistant to being carried away by the air current during the cleaning process.

Brief microwave treatment

When seeds are subjected to a brief exposure of microwaves they respond with enhanced germination. This treatment must be carefully controlled, is appropriate only for small seed lots, and is mainly used to help rescue low germination seed, especially seed of endangered varieties. I have successfully used this treatment on a lot of low-germination pepper seed, using an exposure time of 15 seconds to increase the germination by about 25%. Longer treatment times were not attempted. This method is not recommended for large seed lots because of the difficulty in obtaining equal microwave exposure in all areas of the treated seed lot. The procedure involves placing a thin layer of seed in the center of the microwave oven and microwaving for approximately 10 to 30 seconds depending on the size and water content of the seed.

Sunflower seed, when subjected to 30 seconds of microwave treatment, germinates faster than nonmicrowaved seeds, but when microwaved for 60 seconds they fail to germinate. Microwave radiation selectively heats the water within the seed. In addition, fats and oils are heated, though to a lesser extent. Fava beans microwaved for 30 seconds show a seven-fold increase in certain nutrients. The heating process speeds up metabolism and there is some speculation that the heating may affect the permeability of the lipid-containing membranes, thereby causing cellular changes leading to enhanced germination. If the seed is heated too rapidly or too long, certain biological molecules (proteins) such as enzymes are denatured, losing their enzymatic activity, and the seed is killed.

Seeds that have a low seed moisture content may need longer exposures than seeds with a high seed moisture content. For most seed treatments the initial exposure should be anywhere from 10 to 15 seconds, with additional trials at 10 or 15 second increments. If this procedure is used for germination enhancement, some preliminary experiments should be done to determine the optimum exposure for germination enhancement for a particular type of seed. The effect of microwaves on seed germination is not well documented, so reports of any experiments are welcome and should be forwarded to the author. Because microwaving seed constitutes a seed treatment, check with your organic certifier before planting the seed or using it in commerce. In any case, microwaved seed needs to be retested after treatment.

LABELLING AND RECORD KEEPING

Labeling

During the growing season, through planting, harvest, and

processing it is essential that all crops be properly labeled. This is especially important once the crop is harvested, as many seeds may look alike. Labels should be on every container and should travel through every step of process and cleaning. While this may seem obvious, it is very easy to think that you will remember what was in a particular container, but it is surprisingly easy for several hours to slip by, and then to get confused about the identity of a particular seed lot. Also, when seed is harvested at different times from the same variety, it is also useful to put a date on the label. This can be important later on if there are questions about purity due to improper isolation, disease, or other issues that arise retrospectively.

Record Keeping

You should keep a log of most everything you do with your crops, including information about planting times and location for each variety, lot number and source of seed planted, germination percent or comments about germination and vigor, number of plants transplanted (useful for calculating seed yields later), maturity data, yield data, insect problems, organic fertilizers and amendments, and other data which may become useful later. Your garden or farm log is also important for organic certification. Detailed records should be kept in a card file or computer database regarding variety characteristics and performance. Such data should include an accession number and information about seed source, lot number, year last grown, maturity data, plant characteristics (growth habit, yield, productivity, color and shape of fruit), disease resistance or susceptibility, and taste comparisons with other varieties. Record keeping can be time consuming, so from the perspective of time management it is important to keep in mind that the real value of information is information re-use.

REFERENCES

Bewley, J. D. and M. Black, *Seeds: Physiology of Development and Germination*, Plenum Press, NY, 1985.

Copeland, L. O., *Principles of Seed Science and Technology*, Burgess Publishing Co., Minneapolis, MN, 1976.

Deno, N., *Seed Germination Theory and Practice* (2nd ed.), State College, PA, 1993.

Grabe, D. F. and D. Isely, "Seed storage in moisture-resistant packages" *Seed World* (Jan 24.) p. 4., 1969.

Harrington, J. F., "Drying, storing, and packaging seed to maintain germination, vigor" *Seedman's Digest*, 1960.

_______________ ., "Seed storage and longevity", In Kozlowski, T. T. *Seed Biology*, 1972.

Justin, O. L. and L. N. Bass, "Principles and Practices of Seed Storage", *Agriculture Handbook Number 506*, U.S.D.A. Agricultural Research Service, Washington, DC, 1978.

17

Seed Standards and Certification

The efforts of forest geneticists and tree breeders to introduce superior provenances and improved cultivars into forest practice depend on the organisation of an adequate supply of seed for which special seed producing units are required. Seed of the best provenances is produced in selected parts of natural forests or selected plantations which have been converted into seed stands or seed production areas; while seed of new improved cultivars is produced in seed orchards which consist of grafted plants or plants of seedling origin, derived from selected parent trees.

Besides being of a superior provenance or cultivar, the seed which reaches the forester for sowing in the nursery or in the forest must be clean, viable and healthy if wastefully low yields of plants or crop failures are to be avoided. In more precise terms, the seed must be free from impurities, possess high germinative capacity and be free from seed-borne pests and diseases.

Thus, if the productivity of the forest is to be raised through the use of the products of selection and breeding, the supply of seed and plants must be so organised and regulated that it is possible to obtain seed and plants that are both true-to-name and satisfy certain minimum requirements of quality.

FLOWERING, SEED FORMATION AND SEED PRODUCTION IN FOREST TREES

Seed production areas and seed orchards exist to produce the greatest possible yield of well-filled viable seed. Their successful management depends on a thorough knowledge of the genetics and physiology of flowering and seed production, and it is therefore necessary to summarise the present state of knowledge on these subjects.

Initiation of Flower Buds

In many temperate tree species flower primordia are formed early in the growing season preceding the spring in which the flowers appear. The critical period appears to lie between early May and the end of July. A certain minimum degree of heat is apparently necessary for flower bud initiation. Trees grown in sunny positions come into flower earlier in life and flower more regularly and prolifically than those grown in a close stand or in shade.

A reduction in water supply in summer is frequently associated with flower bud formation but lack of moisture evidently can become a limiting factor and cases are known where irrigation resulted in an increased seed crop. Sarvas has provided ample evidence for Pinus sylvestris that flowering and seed production are better on the more fertile sites.

There is no doubt that on many forest sites the application of fertilisers will stimulate flowering and seed production but the results obtained so far have been very variable. Responses in flowering and fruiting have been obtained from the application of nitrogen, phosphorus and potassium applied singly; from combinations of nitrogen and phosphorus, and phosphorus and potassium; and from complete NPK fertilisers. In at cases where the fertiliser was applied in combination with crown release by thinnings, the combined treat meets produced better results than were obtained with fertiliser alone, and the same obtains where fertilises has been applied in combination with irrigation.

Little is known of the role of plant growth substances and the internal changes in forest trees associated with flowering, but the success of Japanese workers in using gibberellic acid to induce flowering in a number of coniferous tree species encourages further research into the role of plant growth substances in flower initiation.

Flowering and Pollination

The importance of adequate pollination for the development of viable seed by forest trees has been well demonstrated for several species. Cross pollination is the general but by no means universal rule in many tree species of economic importance, and a considerable number are more or less self-sterile.

It now appears that the dispersal of pollen in wind-pollinated species is not so dependent on the weather prevailing during the period of flowering as was once thought; the reason appears to lie in the great speed and quantity of pollen dispersal as soon as conditions are right. Extensive observations on the flowering of Betula verrucosa, B. pubescens, Picea abies and Pinus sylvestris in Finland led Sarvas to conclude that the start of flowering, the duration, and the occurrence of the peak of flowering are all closely dependent on temperature. Strong winds may accelerate pollen release and rain may interrupt it, though rain seems to have less effect on pollination and seed set than has often been assumed.

In the case of insect-pollinated species, such as Acer pseudoplatanus and Tilia, good flying weather for insects is required for pollination, that is bright, dry and fairly calm weather. The reader is referred to the paper by Sarvas for information on the design of pollen catching equipment.

There are many records of small amounts of airborne tree pollen traveling great distances from the source ("foreign pollen ") but several workers have shown that the bulk of airborne pollen falls fairly close to its source (" local pollen "). The results suggest that pollination is mainly effected by trees in the immediate vicinity of the seed tree. However, the behaviour of pollen produced by stands of trees must also be considered as well as factors affecting the distance of dispersal including the size and shape of the stand, and the quantity and nature of the pollen produced. Pollination conditions are best in extensive stands containing trees with well-developed crowns.

Fertilisation and Seed Formation

Pollination is normally followed by fertilisation. The process of pollen formation, pollen shedding and sub sequent germination on the tip of the nucellus in Larix and Pseudotsuga is described in detail by Barner and Christianson. The lapse of time between pollination and fertilisation is usually quite short, but in Pinus the interval is about one year.

The comprehensive study made by Sarvas can again be referred to for a good account of the normal course of fertilisation in Pinus sylvestris. The pollen grains are normally very viable, they develop sperm nuclei with great regularity and these are carried with great accuracy to the eggs by the pollen. The eggs are generally fertile and there appear to be no genetic barriers that significantly prevent fertilisation. However, the competition between developing embryos is often decided by genetic factors as, for example, when the union of semi-lethal genes to form homozygotes at fertilisation leads to the defeat of the embryo during the competition in the embryo cavity.

Nonpollinated ovules gradually collapse, mostly during the first growing season and in Pinus poorly-pollinated cones soon drop. High embryo mortality in

Pinus sylvestris is attributed by Sarvas to self-fertilisation, and Ehrenberg et al. also showed that self-fertilisation often results in poor germination of full seeds and inferior progenies.

Size of the Seed Crop

Trees that bear heavily one year and then sparsely or not at all for several years are said to show periodicity in seed-bearing. In many species the seed crop of one year affects the seed crop of the following year, both through the reduction of potential foliage-producing or vegetative buds and the reduction of carbohydrate resources. The flowering and seed production of young trees is commonly sparse and sporadic but increases rapidly with increasing age and size. Within a forest stand it is the larger, dominant trees that bear the most seed but there are also inherent differences between trees in flowering, fruiting and the production of viable seed that make necessary careful selection for fertility and fruitfulness.

Good flower years occur much more frequently than good seed years. There may be several reasons why a spate of flowers often fails to produce a spate of seed, but deficient pollination, the effects of self-fertilisation, the activities of seed-destroying and cone and fruit-damaging insects, animals and birds and unfavorable weather are usually to blame for this. Periodicity in flowering and seed production depends to a large extent on external factors and protection from climatic dangers and the animals, birds, insects and fungi that damage or destroy flowers, fruit and seed should do much to mitigate, but not prevent irregular seed production. For many species the gap between good seed years can be bridged by the use of refrigerated seed storage.

Quality of the Seed Crop

The spells of warm, dry sunny weather which appear to

favor the initiation of flower buds and the accumulation of carbohydrate resources also favor the maturing of the fruit and seed. Considerable variation in seed quality and time of maturation has been found between individual trees and stands of the same species.

In northern latitudes a greater part of the cone or fruit crop is borne on the south than on the north side of the crowns of seed trees. This distribution affects the methods used to assess the size of seed crops. In many species the best cones, fruits and seeds are generally found on the middle and upper parts of the crowns. There is also a variation in seed quality within the cones of many coniferous species, the middle portion being the most productive of well-fired viable seed.

Two general recommendations for the improvement of seed supplies can be made.

— Full advantage should be taken of good seed years in all species, because the highest proportion of well-filled viable seed is produced in such years.
— Much can be gained by improving seed collection methods both from standing trees and from the ground as in Fagus, Quercus and Tectona.

TREATMENT OF SEED STANDS

Objects and Choice of Seed Stands

The objects of forming seed stands or seed production areas are to:

— produce seed of improved inherent quality by selecting and favoring seed trees which are vigorous, straight-stemmed and healthy and produce wood of good quality;
— concentrate seed collection into a few specially treated parts of the forest, thus making seed collection easier to organise and control;

— improve the germinative energy and germinative capacity of the seed collected.

All three objects can be achieved by one simple treatment, that is, the careful selection of seed trees and the complete isolation of the crowns of these seed trees by thinnings.

A necessary preliminary to the formation of seed stands is a survey of the region to locate suitable natural stands or plantations. A seed source is a crop of fast-growing and healthy trees with good growth habit suitable for seed production. Plus seed sources are suitable for regular and intensive seed collection; normal seed sources are suitable provided careful selection is made of the seed trees; minus stands are unsuitable for seed collection. A seed source becomes a seed stand or seed production area when all defective trees have been removed, leaving the best trees isolated to develop big crowns and so produce more seed. The ground and the seed trees may also be treated to increase the seed crops and make collection easier.

Seed stands are usually formed in natural forests or plantations that are sufficiently old or sufficiently well-developed to provide reasonable assurance that they are well adapted to the site and will continue to show rapid and healthy growth. A suitable lower limit is a top height of 11 to 12 meters (36 to 39 feet) and a good upper limit is half the rotation age.

Clearly it is desirable to convert relatively young stands because the response of younger trees to thinnings and fertilisers is greatest and there will be opportunity as the trees develop in later years to make further selection among the seed trees. Nevertheless, many outstanding or plus stands approaching or at the age of maturity for timber purposes can form valuable seed stands. The area of seed stand required can be fixed by observing the average seed production of well-grown trees and setting this against the future planting or regeneration program.

Isolation from Foreign Pollen

An additional criterion in choosing seed stands is the absence of neighboring natural stands or plantations of the same or closely related species which are of poor or minus quality. Such isolation from inferior sources of pollen is very important.

Nevertheless, when fixing the size of isolation strips around the seed stand and isolation distances from large sources of undesired foreign pollen, it should be remembered that the degree of contamination by foreign pollen is roughly proportional to the size of its source, and that contamination is least when the size of the seed stand and the local pollen cloud are large in relation to the cloud of foreign pollen. These considerations are the basis for the recommendation that seed stands should be at least 3 and preferably 5 hectares (7.5 to 12 acres) in extent.

Selection of Seed Trees

The first step in forming a seed stand is to select and mark the seed trees. These should be fast-grown healthy dominants capable of producing wood of good quality. The stems should be persistent and straight, and also free from defects such as fluting, spiral grain and epicormic shoots. The branches should be smart in relation to the stem at the point of origin, the branch angle should be flattened to moderately ascending, and good natural pruning is also a desirable feature. The crowns should be compact and well-provided with foliage and the trees should show signs of having borne seed in the past. Where wood characters such as specific gravity or tracheid and fiber dimensions can be clearly specified, it may also be possible to select seed trees for these characters.

The number and relative superiority of the seed trees marked have an important effect on the genetic improvement of the progeny raised from the seed collected in the seed stand. More precisely, the genetic gain or change in the average genotype of the population produced by the selection of seed trees depends on the heritability of the characters for which selection is made, the variability of these characters and the proportion of the population selected. The genetic gain from seed stands is greater when the proportion selected is low, but the number of seed trees per hectare must be sufficient to ensure adequate cross-pollination between the seed trees, thus avoiding the effects of self-pollination or inbreeding.

Thinning Seed Stands

The second stage in forming a seed stand is the removal of the phenotypically inferior trees to free completely the crowns of the seed trees. Sometimes seed stands are formed from groups or avenues and such trees often have long, well-developed crowns. In most cases, however, the seed stands will previously have been managed for wood production so that the crowns of the seed trees are relatively small.

This is particularly the case in the light-demanding species of Pinus and Larix and also holds for Tectona grandis, Eucalyptus species and Casuarina. The frequency and intensity of thinnings to achieve full crown release must depend on local conditions, and particularly the danger of windthrow, but the sooner the seed trees are isolated the better for future seed production. The first thinnings should be in the nature of heavy crown thinnings to free the crowns of the seed trees, and to remove large, rough dominants from the seed stand and isolation strip. Thereafter the remaining trees are removed as soon as convenient or safe.

Fertilizer Treatments

The next step is to consider what other treatments should be applied to increase flowering and seed production still further and to make their occurrence more regular. Root pruning and stem girdling or strangulation are likely to be more harmful then helpful, but the application of fertilisers stands out as a beneficial and worthwhile treatment, especially where soil fertility is relatively low.

The quantities and relative proportions of the various major elements which are necessary to stimulate flowering and seed production obviously vary with the nature of the crop, especially the age and size of the trees and the site conditions. It is, however, clear that the fertiliser should correct existing deficiencies and supplement the major nutrients and for this a general prescription such as 2N: 1P: 2K at the rate per hectare of 112 kilograms of nitrogen, 56 kilograms of phosphorus and 112 kilograms of potassium (or about 100 pounds of nitrogen, 50 pounds of phosphorus and 100 pounds of potassium per acre) would appear to be a good starting point for fertiliser treatments. Usually the fertiliser is applied over the whole area but where the seed trees are scattered it is preferable to apply it below each seed tree to an area of ground one-and-a-half times the diameter of the crown.

In general the best time to apply compound fertilisers is in early spring before the new flower buds are differentiated: the period February to May is suitable in the temperate regions. But the most effective time of application does vary somewhat with the major elements, and it appears that nitrogen should be applied in late spring but not later than May. Potassium can also be applied in spring but phosphorus may best be applied in autumn. Finally, it appears certain that fertiliser application will in most cases have to be repeated if high seed production is to be sustained. Irrigation cannot be dealt with in detail here but it may have an important part to play in areas where water supply can become limiting to flowering and seed production.

Other Treatments of Trees and Ground

The stems of seed trees in older seed stands should be high-pruned to the base of the living crown to facilitate the use of the Swiss tree bicycle or special ladders for climbing to collect the seed, but no live branches should be removed from the base of the crown because this is the region of pollen production in many species. If truck-mounted ladders are used, any undergrowth and lop and top will either have to be removed from the site or stacked in heaps so that the truck can be maneuvered.

Strong weed growth often develops on the forest floor following the application of fertilisers and heavy thinnings; if allowed to grow unchecked, shrubs and weeds will be a serious barrier to the collection of the fruits and nuts of Quercus, Fagus and other species which are picked from the ground. Such weed growth can be suppressed by regular cutting and clearing or by the use of complete weedkillers such as "Simazine" or "Monuron" which have given promising results in apple orchards. Another possibility worth serious consideration is the use of leguminous ground crops; this has been the subject of much investigation for maintaining the health and yield of Hevea plantations in Malaya.

Where there is little or no ground cover, the fallen seed can be easily seen and taken by birds and the use of bird-scaring devices may become necessary.

Protection Against Fungi, Insects and Animals

The effects of these agencies in reducing flower, cone and fruit crops have already been stressed. The immediate application of creosote or other recommended chemical to the freshly cut stumps of felled trees in coniferous seed stands will reduce the spread of infection by Fomes annosus. Hardwood stumps which show signs of attack by Armillaria mellea should be grubbed up and burned where this is practicable.

When crown feeding insects assume epidemic proportions, as, for example, Bupalus piniarius on Pinus sylvestris, control should generally be achieved by applications of DDT, using ground fogging machines; aerial sprays are not normally economically justified. For tall trees in seed stands and smaller seed trees in seed orchards, systemic insecticides may eventually provide a suitable means of protection against the attacks of shoot borers and Adelges species, but this method is still at an early stage of development. Of the insects which attack flowers and seeds, Megastigmus species are among the most important at present.

The gray squirrel (Sciurus carolinensis) is a serious pest feeding on nuts, buds and bark. It can be controlled by poking out drays and shooting or trapping. Wide metal bands placed round the stems of widely spaced seed trees have been used in the United States of America to prevent red squirrels climbing.

Many species of both wild and game birds eat fallen seeds and some, for example the crossbill (Loxia curvirostra), extract seeds from cones. Control measures may be necessary and will depend on the wild life laws of the country concerned.

TREATMENT OF SEED ORCHARDS

Zobel et al. have defined a seed orchard as a plantation of genetically improved trees, isolated to reduce pollination from genetically inferior outside sources and intensively managed to produce frequent, abundant and easily harvested seed crops. It is established by setting out clones or seedling progeny of trees selected for the desired characters. Seed orchards are often established while genetic evaluation of the parent material is still under way. In such cases components shown by progeny testing to be genetically undesirable will subsequently be removed from the seed orchards.

Types of Seed Orchard

There are several kinds of seed orchard. Clonal seed orchards consist of grafts, rooted cuttings or layers; while seedling seed orchards are composed of selected progenies derived from open-pollination or controlled crosses. Clonal seed orchards may be subdivided into those in which the clones are from plus or untested parent trees; and those composed of clones from elite or tested parent trees.

Concentrating first on clonal seed orchards based on plus trees, three kinds may be recognised on the basis of the origin of the clones:

— Plus trees are selected from a specified geographic or climatic region; or from a group of stands; or, more rarely, from a single stand. All the plus trees are of the same species.

— Plus trees are of the same species but originate from different geographic regions. These are called "provenance seed orchards" by Andersson who described one such seed orchard containing Swedish and Polish clones of Picea abies.

— Plus trees are of different species and the object is to produce interspecific hybrids. Examples are the seed orchards planted to produce seed of the F1 hybrid Larix decidua x L. leptolepis = Larix x eurolepis.

Clonal seed orchards composed of elite or tested clones will form the next generation of seed orchards. Their formation is dependent on systematic test crosses made in the seed orchards composed of plus or untested clones.

Seedling seed orchards have been advocated by Wright who bases his argument on a theoretical comparison of the response to selection or genetic gain to be expected from several alternative selection procedures.

Test Crosses in Seed Orchards

The design of test crosses and the associated progeny

testing have already been discussed in Chapter 2. To repeat the main point, the object is to identify clones showing low combining ability so that they may be removed from the seed orchard. The minimum number of clones in seed orchards composed of plus trees is now generally accepted to be 20 or 30, and a common maximum number is 60 clones. When elite or tested clones are available 2-clone orchards are envisaged by Andersson.

Choice of Site and Ground Treatments

Seed orchards are planted on sites well isolated from large areas of plantations of the same or related species, the distances recommended being up to 1,000 meters (1,100 yards), and favorable to the regular initiation of flower buds and the regular production of large crops of well-filled viable seed. The local climatic requirements are shelter from strong wind and adequate air drainage to reduce the damage done to flowers and fruits by frosts. Soils with a good nutrient status are preferred; the evidence presented in the first part of this chapter suggests that is it generally a mistake to site seed orchards on soils of low nutrient status.

There are three possible ground treatments during the early growth of seed orchards: clean cultivation (this being the treatment favored in Sweden at the present time), cover crops and sward. Once the seed trees are well established, the sward appears to be a convenient treatment for many temperate tree species because of the relatively simple management and, by analogy with fruit trees, beneficial effects on the potassium and phosphorus status of the seed trees.

On the three main types of sward used in fruit orchards, that is, the "tumble down," white clover sward and grass-white clover sward, the last appears to satisfy the three conditions of contributing soil humus, supplying some nitrogen (through the clover) and providing good ground cover. It should be noted, however, that drought effects can develop under uncut sward conditions so regular grass cutting or grazing is essential. Again, as with seed stands, grass swards are not suitable for all species and leguminous cover crops are worth consideration.

Manurial regimes for seed orchards are unlikely to differ much from those already suggested for seed stands, with the possible exception of methods of application. In top-fruit orchards (that is, apples, pears, etc.) certain fertilizers, in particular nitrogen and magnesium, are often more readily absorbed by the leaves of trees than through the roots. Olroyd made special reference to foliar applications of urea to top fruits.

Distance between Seed Trees

The most suitable distance between seed trees varies with species and methods of management, but initial planting distances should not be too wide because pollen production will be sparse and consequently the yield of viable seed will be rather low in the early years. Moreover, at a later stage, clones or seedlings shown by test crosses and progeny tests to be genetically inferior are also removed from the seed orchard, thus giving more growing space for the trees which remain.

Layouts designed to allow one or two "mechanical" thinnings are now general. The distances recommended range from initial spacings of 3 meters (10 feet) for Pinus sylvestris and Fagus sylvatica to final spacings of 6 to 9 meters (20 to 30 feet) for species of southern American pines. The arrangement of the clones of grafted plants or seedling progenies must meet the requirements of adequate cross-pollination and reduced self-pollination and so a randomised distribution of single seed trees is general. The value of clone or tree banks as a means of preliminary screening of clones for time and pattern of flowering, self compatibility and other related characters is worth stressing here.

Treatment of Seed Trees

In seed orchards the seedlings, grafts or cuttings should be encouraged by careful choice of site and favorable treatment at planting to grow rapidly from the beginning; this should lead to an early onset of flowering and fruiting. Thereafter, everything should be done to favor rapid growth and the best treatments appear to be regular crown release, the application of organic and inorganic fertilisers, control of the ground cover, pruning and geotropic treatments and, in some areas, irrigation. It may also be necessary to stimulate flowering by excessive applications of fertilisers, restricting water supply, stem girdling or root pruning, but in general these should be regarded as emergency or experimental, rather than regular, treatments. In the long term, choice of rootstock and treatment with plant growth substances will play their part.

Shoot Bending, Pruning and Other Crown Treatments

The orientation of the shoot has a marked effect on flower initiation in apple and species of Larix. The geotropic or branch bending treatment developed for Larix by Longman and Wareing can be applied in larch seed orchards by tying branches of the seed trees into a downward pointing position during the dormant season. Disbudding to induce or increase flowering has proved useful in Pinus sylvestris, and pinching off the shoot tips before meiosis takes place has been used in Japan in experiments on the artificial control of sex differentiation in Pinus densiflora and P. thunbergii. There is also scope for much more work on the effects of foliar application of growth regulators and nutrients on the initiation and differentiation of flower buds. Finally, it is clear that attention should in general be focused on the use of shoot bending, pruning, nutrients and growth substances to increase pollen production, which is often insufficient in young seed orchards of Pinus species.

Selected Rootstocks in Grafting

It is now commonplace to find that grafted plants, formed by grafting scions taken from adult flowering trees on to seedling rootstocks in the juvenile nonflowering condition, soon begin to flower and bear seed. This is the basis of seed production in clonal seed orchards. In fruit orchards, selected clonal rootstocks are used to control the size of the grafted tree and the nature of the fruit bearing, and it would appear that the use of selected clonal rootstocks will also lead to similar benefits or uniformity and control of tree behavior in tree seed orchards.

The idea is not new, Johnsson and others having grafted Pinus sylvestris scions on to seedling rootstocks of Pinus mugo and two origins of Pinus sylvestris. Analysis of the cone and seed harvest of 1959 and 1960 from these trees failed to reveal any influence of the rootstocks on the seed production of the scions, so the only evidence so far is negative. Nevertheless, an additional problem associated with the interaction of rootstock and scion is that of incompatibility of the graft union and this has been reported for Pseudotsuga taxifolia and Fagus sylvatica and may make necessary increased attention to selection of rootstocks.

Protection Against Fungi, Insects and Animals

A factor which may have to be considered in the siting and subsequent treatment of Pinus seed orchards is the damage done by cone rusts. An example is provided by Cronartium strobilinum which damages the cones of Pinus elliottii and P. palustris. In young seed orchards of Larix, Pinus and Pseudotsuga taxifolia the Adelges species which commonly appear on the grafted plants can be controlled by "Malathion" sprays. As in seed stands, systemic insecticides may also prove useful as a control measure in the future.

CERTIFICATION OF FOREST TREE SEED AND PLANTS

The object of the certification of tree seed and plants is to maintain and make available to the practicing forester sources of seeds, plants and other propagating materials of superior provenances and cultivars so grown and distributed as to insure the genetic identity and high quality of the seed and plants.

Superior provenances are identified by means of provenance tests. The duration of these tests depends on the rate of growth and the length of the economic rotation for each species but, in general, it must be accepted that provenance tests are essentially long-term experiments. The need to provide safe prescriptions for the movement of seed and plants so that expensive losses in wood production can be avoided has led foresters in countries with large variations of site and climate to devise zoning schemes defined in terms of latitude or distance and elevation.

Thus in Austria, Czechoslovakia, Norway, Sweden and the Federal Republic of Germany to name a few such countries, seed is collected and the plants are utilised within the same zone. Countries with smaller ranges of site and climate tend, when importing seed, to look for seed sources growing in places with similar features of site and climate, although as the results of provenance tests accrue the specification can sometimes be made less strict and the search for superior provenances tends to range farther afield.

The state forest services of many countries maintain registers of seed sources which list classified seed sources for home use and export, and acceptable provenances that can be imported. General lists of acceptable provenances covering a wide range of species have been published in the

United Kingdom, Denmark and Belgium and more restricted lists are available in other countries. Even if the best provenances are not yet known, foresters are often clear about what they should not use.

The rapid progress made with selection and breeding in the genus Populus has been largely due to the easy propagation of many species and hybrids by means of stem cuttings. There is a very wide range of distinct and stable cultivars in use, and control of genetic identity is exercised in most western European countries, partly by control of the sales of planting stock (as in Belgium, the Netherlands and Spain) and partly by the distribution of cuttings from registered stool beds (as in Belgium, France, Italy, the Netherlands and the United Kingdom).

The production of cultivars of those species of forest trees that are reproduced through seed is now developing rapidly and, as the seed production areas and seed orchards described earlier in this chapter become more productive, an increasing quantity of seed and plants of superior provenances and improved cultivars will be available for general use. The efforts of forest geneticists and tree breeders will be largely nullified unless certification procedures are available to maintain the genetic purity and quality of the seed and plants.

Categories of Seed and Plants

Four broad categories of tree seed and plants are moving in national and international trade and these categories are distinguished by the precision with which origin is described; the degree of selection of the seed trees and their isolation from foreign pollen; whether the offspring of the seed trees have been tested or not; the amount of supervision and independent inspection of seed collection, processing and storage, and plant production; and the standards of record keeping and labeling. It is convenient to describe these four categories of seed and plants in ascending order of genetic value.

— *Unclassified seed and plants.* The origin of this category is unknown or inadequately described, or the seed has been collected in stands not classed as seed sources by a recognised agency because of poor quality of the trees or lack of isolation from inferior trees of the same or closely related species. Unclassified seed is largely collected from felled or stunted trees and the seed, and plants are collected, raised and marketed with little or no supervision. The object of every country is to discard this category of seed and plants as rapidly as possible.

— *Source-identified seed and plants.* This category is derived from good natural stands and plantations registered as seed sources by a competent service. The place or region where the seed is collected is clearly defined; the seed is harvested, processed and stored and the plants are raised under supervision and independent inspection; the labeling is adequate and records are available for inspection.

— *Selected seed and plants.* The seed is collected from carefully selected seed trees growing in natural stands and plantations classified as better than average and registered by a competent service. Trees that do not conform to stated standards of growth rate, stem and crown form, branching habit, health and other characters are removed as rapidly as possible from within the stand and from an isolation surround of prescribed width. These seed stands or seed production areas are also treated to increase seed production by means of fertiliser applications and protective measures against the animals, insects and fungi that damage seed. The statements made for source-identified seed about supervision, independent inspection, labeling and records also apply to selected seed.

— *Certified seed and plants*. This category is derived mainly from clonal seed trees in seed orchards (and also from elite trees and stands) whose genetic superiority has been proved by progeny testing to standards defined by a competent service: Certified seed and plants are produced in a manner that insures genetic identity and proved interspecific hybrids may be included in this category. Clonal seed orchards contain highly selected clones arranged to favor cross-pollination. The site is well isolated from pollen of inferior trees of the same or closely related species. Certified seed is harvested, processed and stored, and certified plants are raised under supervision and independent inspection. The labeling is adequate and records are available for inspection. In particular, the exact sources of the individual components of the seed orchards and the results of progeny testing must be readily available. Seed collected from seed orchards before progeny testing is completed is sometimes sold as plus or selected seed.

Administrative Control of Seed and Plant Production

It is believed that there are 12 comprehensive national certification schemes and these can be grouped into 3 classes: voluntary schemes operated by independent seed associations; voluntary schemes operated by state forest services; and compulsory schemes operated by state forest services. The Czechoslovakian and Polish procedures are based on virtual state monopoly of tree seed and plant production, both for internal use and export. In the other countries with certification schemes, particularly the United States and central and northern Europe, there is a strong private seed and nursery trade.

In the Federal Republic of Germany the Federal Seed Act of 1957 provides legal backing for a full official certification service. The scheme used in Georgia, U.S.A., is backed by two state laws but is voluntary; it is based on widely accepted procedures sponsored in the United States by the International Crop Improvement Association. In Finland and the Netherlands, there are private tree breeding associations that gain a good part of their revenue from the sale of certified seed and plants.

Those certifying agencies operated by independent associations (as, for example, those in Denmark and the United Kingdom) are directed by a seed board or management committee elected by and representing the seed and nursery trade, forest owners, the state forest service and forest research institutes (including the seed testing agency). An important principle is that the purchasers of seed and plants should be represented. The management committee decides policy, sets up rules governing the certification program and has a professional forester as executive secretary or manager.

The voluntary certification agencies operated by state forest services usually have an officially-appointed supervisor who is responsible for the certification program under the direction of an appointed board that determines policy and sets up rules. Compulsory certification schemes are based on seed laws and are operated by the state agencies named under the seed laws.

Most certification schemes are nonprofit-making and largely financed by fees paid by the seed producers and nursery trade. These fees usually include a forest or nursery charge and an area inspection fee. Additional charges may be levied by means of official stamps or labels provided to the merchant by the certifying agency at a predetermined unit cost. The selling price of seed and plants of the superior categories is increased, often considerably, but the cost of seed is a very small fraction of total establishment costs and it can be shown that small increases in yield justify large increases in seed prices.

Outline National Certification

The elements of a comprehensive national certification scheme for forest seed and plants are as follows:

— Inspection of the seed source by a qualified professional forester before pollination of the seed crop. At this time (which is from 6 to 21 months before seed collection) the quality of the seed trees, incidence of inferior trees and effectiveness of the isolation can be checked. Many forest trees do not produce seed every year and, after the first inspection, subsequent re-inspections are made only in years when seed is to be collected (the longest interval being about 5 years).

— Assessment, if possible, of the cone or fruit crop by a qualified professional forester at a stated time (about 90 days) before collection begins.

— Collection of the cones of fruits by a registered seed collector; extraction, cleaning and packaging of the seed at a registered seed extraction plant; and storage of the seed at a registered seed store. The records of collection, processing and storage are made available for inspection and the labels must conform to minimum requirements.

— Testing, under the rules of the International Seed Testing Association, of an adequate sample of the seed at an official seed testing station.

— Sowing of the seed in a registered nursery where labels and records satisfy minimum requirements. Inspection of the seedlings and transplants should be made by a qualified professional forester before they are lifted and dispatched.

Although growing-on tests are not yet widely used, the progress of research in several countries holds promise of their eventual use on a wide scale.

Minimum Requirements for Certification of Seed and Plants

The prescriptions that follow have been derived from the published accounts of national certification schemes in 12 countries. The prescriptions also incorporate relevant details from the scheme for the varietal certification of herbage seed sponsored by the Organisation for Economic Co-operation and Development.

Previous history of the seed source. A seed stand or seed production area may be formed in a natural stand (established by natural regeneration) or plantation (established by planting or sowing) of indigenous, introduced or unknown origin. In some countries, such as Czechoslovakia, France, the Federal Republic of Germany, Poland and Switzerland, determined efforts are being made to preserve and use the indigenous growing stock.

In other countries the replacement of the indigenous growing stock by introduced species and provenances is very advanced and the exotic material must be accepted. Finally, some countries exclude from their certification schemes any stands of unknown origin, but although this is desirable it is not always practicable.

Isolation from foreign pollen. The requirements for isolation of seed production areas and seed orchards from large stands of inferior trees of the same or closely related species and for the width of isolation strips vary somewhat with species and country, but a generalised prescription for seed production areas and seed orchards is: 1,000 meters (1,100 yards) from a large stand of inferior trees and 100 to 150 meters (110 to 165 yards) from the nearest inferior tree - thus fixing the width of the isolation strip.

Field inspection of seed trees, seed crop and plants: Inspection of seed sources, seed production areas and seed orchards is most often made by the state forest service or by qualified professional foresters nominated by the certifying agency. The standards for seed production areas are based on the number of vigorous, well-formed and healthy seed trees per hectare and removal of inferior trees. Inferior trees are marked by the inspector and if possible felled while he is in the area, but the rate at which inferior trees are removed is often determined by local silvicultural requirements and re-inspection may be necessary when the fellings are completed.

During seed collection and processing, the cones, fruit and seed must be handled so as to prevent mixture and maintain identity. Similarly, mixture must be prevented and identity maintained in the seed store and when raising plants in the nursery.

Standards for germination and purity. The methods of determining purity, germination and moisture content and the standards of health supervision with regard to seed-borne pests and diseases are generally based on those set by the Forest Tree Seeds Committee of the International Seed Testing Association (ISTA) Standards of germination and purity for purposes of sowing the seed and getting a good yield of plants are available for a very wide range of species.

Labels and certificates. It is generally agreed that the amount of information required on labels should be kept within reasonable bounds, and a number of distinctive yet simple labels are in use. Most of these bear the following items of information:

— Species (Latin name)
— Subspecies variety or cultivar name
— Category (source-identified, selected, certified)
— Provenance (reference number, place or region, elevation)
— Supplier (name and address)
— Seed (month and year of crop, quantity)
— Plants (age or size, quantity).

In 1950, FAO and IUFRO jointly prepared an international certificate of origin and quality for tree seed and plants and this carries a large amount of information. This certificate has great value in exchanges of small quantities of seed for provenance research, but something simpler is needed for bulk trading in tree seed and plants. Such a certificate would bear the following items:

— Certifying agency
— Consignee (name and address)
— Sender (name and address)
— Species (latin name)
— Subspecies variety or cultivar name
— Category (source-identified, selected, certified)
— Provenance (reference number, place or region, elevation)
— Gross weight of packages
— Contents (seed or plants, quantity)
— Disinfection treatment (date, place, method).

Reference numbers. Comprehensive yet simple methods of identifying the origin of seed and plants by means of reference numbers or letters are of great assistance in administering a comprehensive certification scheme. FAO has adopted the universal decimal classification system by which each country is identified by a permanent number consisting of three digits.

Certification of Forest Tree Seed and Plants Moving in International Trade

Rohmeder points out that tree seed moves in national or international trade for three main reasons:

— in large quantities for large-scale cultivation of a species, provenance or cultivar, as, for example, in Pseudotsuga taxifolia, Pinus radiata, Tectona grandis and Eucalyptus;

— in small quantities for research purposes, such as species trials, provenance tests, progeny tests and tests of cultivars produced by selection and breeding;

— as an item of food, such as chestnuts (Castanea sativa) walnuts (Juglans regia) or olives (Olea europaea).

The discussion that follows deals entirely with the first situation and is intended to refer to both seed and plants.

In Europe there has been regular and extensive international trade in seed and plants of Pinus sylvestris, Picea babies, Larix decidua and Abies alba for a very long time. Now there is extensive trading in tree seed between North America and Europe in the coniferous and broadleaved trees of the northern temperate zone, important examples being Pinus contorta, Pinus nigra, Picea sitchensis and Pseudotsuga taxifolia.

Looking further afield, the Eucalyptus species and subtropical pines are playing a rapidly increasing part in afforestation in the Mediterranean region, central and southern Africa, South America, and the Asia and Pacific region. Finally, two tropical species deserve mention, namely the Honduras mahogany (Swietenia macrophylla) and teak (Tectona grandis), seed of which is moving extensively between tropical countries of Africa and Asia. For phytosanitary reasons, the movement of tree seed is more extensive than the movement of plants.

Six international bodies are active in matters affecting trading in forest tree seed and plants:

— The Forest Seeds Committee of the International Seed Testing Association have developed rules for testing the quality of tree seed.

— Section 22 of the International Union of Forest Research Organisations continues to sponsor international provenance research.

— FAO has consistently sponsored improved records for use in the international exchange of tree seed, and organised the World Consultation on Forest Genetics and Tree Improvement to foster the use of genetically superior tree seed and plants.

— The International Crop Improvement Association in the United States has supported specific standards for certifying the origin of forest tree seed, drawn up by the Society of American Foresters.

— Foresters of countries in the European Economic Community have agreed rules for trading in forest tree seed and plants.

— The Organisation for Economic Co-operation and Development (OECD) has sponsored schemes for the certification of agricultural seeds.

The work of these international organisations suggests that the time is coming when internationally acceptable seed certification standards can be framed for forest tree seed and plants. The first steps have recently been taken by OECD which has provided funds to enable a working group to prepare the necessary draft rules and directives for consideration by all member countries.

REFERENCES

Andersson, E., "Seed stands and seed orchards in the breeding of conifers", *FAO/FORGEN*, 63-8/1, 1963.

Anonymous, *Weed control*, 2nd ed. Oxford, Blackwell, 1960.

Astinet, H., *Forest tree seed - specification of provenance*, Saatgut-Wirtschaft, World Seed Campaign Edition, 1961.

Barber, J. C. & Darby, S. P., "Georgia's seed certification program", *Journal of Forest*, 1959.

International Crop Improvement Association, "Minimum seed certification standards", Publ. No. 19. International Crop Improvement Association, 1959.

Isaac, L. A., "Problems and proposals for international certification of tree seed origin and stand quality", Proc. 5th World For. Congr., Seattle, U.S.A., 1960.

Koospol, *Seed and plants of forest tree species in Czechoslovakia*, Koospol, Prague, 1961.

Kozlowski, T. T., "Physiological implications in tree improvement", *FAO/FORGEN*, 1963.

Matthews, J. D., "Factors affecting the production of seed by forest trees", *Forest Abstract.*, 24 (1): i-xiii, 1963.

Paged, *Forest tree seeds and trees*, Warsaw, Paged Foreign Trade Enterprise Export and Import, 1962.

Seal, D. T., "Collection of cones from standing trees", *For. Comm. For. Rec. No. 39*. London, H.M.S.O, 1959.

Tulstrup, N. P., *International trade in forest tree seed*, Unasylva, 13 (4): 196-201, 1959.

Zobel, B. J. et al., "Seed orchards - their concept and management", *Journal of Forest*, 1958.

18

Management of Seed Orchards

Some species reproduce after few years while others await several decades before they set flowers and fruits. Plants will often reproduce late or not at all if they grow (or are grown) on the periphery of their geographical distribution or under adverse growth conditions. On the other hand, the physiological maturity can be hastened if the trees are grown under favorable conditions. For some species with a late reproductive age, the reproduction can be hastened by using grafts or cuttings since these vegetative materials tend to retain their physiological age and consequently flower in an earlier age than plants raised from seeds. The position of the tree from where the scions are taken often influences the flowering habit, too.

PHYSIOLOGICAL AND ENVIRONMENTAL CONDITIONS

Good growth conditions in general also promote good flowering. An ample flowering is a precondition for abundant seed production. In addition to good growth conditions in terms of soil and climate, flowering/fruiting also requires excess light and effective pollination. The main ecological factors that promote flowering/fruiting are the following:

- Optimal soil conditions (pH, fertility, structure, drainage, etc.)

— Optimal climate (water, light, temperature, etc.)
— Optimal conditions for pollination (pollination agent, synchronized flowering, etc.)
— Reduced root competition (for water and nutrients).
— Reduced crown competition (for light).

Internal factors within the plant also influence flowering and seed production. Many trees do not produce a heavy seed crop every year. An abundant seed crop usually occurs when a period of accumulating nutrient resources in the trees is followed by a (brief) physiological stress period, which may be environmental or man-imposed. Following physiological factors promote heavy fruiting:

— High level of internal resources in the plant.
— Effective pollination (compatibility, timing).
— Necessary level of flower promoting hormones.
— Brief stress condition during flower differentiation.
— Optimum nutrient and water supply during fruit development.

Some of the above factors are site specific and must be taken into some consideration when allocating the site for the seed orchards. Others can be manipulated through proper management practice.

Establishment of Seed Orchards

Seed orchards are stands planted especially for the production of abundant superior seeds. A seed orchard consists of trees where the phenotype is usually of minor importance as long as the trees will produce seeds. Only when the seedling seed orchard is combined with a genetic test (cum progeny trial) are the phenotypic characters of the trees important.

There are two main types of seed orchards, named according to the way of establishment:

— Clonal Seed Orchard: Seed orchard raised from selected clones propagated by grafting, cutting, air-layering or tissue culture.
— Seedling Seed Orchard: Seed orchard raised from seedlings produced from selected parents through natural or controlled pollination.

Which type of orchard is best in the particular situation is dependent of, e.g., species and the breeding programme. Some characteristics of the two types are listed below: Under the circumstances where a genetic test/progeny test is converted into a seedling seed orchard, these are the following constraints in terms of locations and management:

— Genetic tests are carried out in areas similar to the potential planting area of future plantation programmes. This is not necessarily conducive for flower and seed production.
— Genetic tests are carried out at narrower spacing with strong competition whereas seed production requires wide spacing.

Size and Longevity

The size of the seed orchard is determined by the seed demand and the expected seed production from the orchard. Both figures are subject to estimates. Planting targets may be altered in the future and the seed production from individual trees are often only subject to qualified guesses. A minimum of 15-25 clones or families to assure a sufficient genetic base and limit selfing should be used.

A seed orchard can in theory be operative as long as the trees are physiologically capable of producing seeds. In practice, however, a seed orchard will only be operative until a new improved seed orchard has been established, based on next generation progeny test. Seed orchards are usually referred to by generation, 1st, 2nd, 3rd, etc.

Location

Since the seed orchard is planted entirely for the production of seeds (with the exception of the combined genetic tests/ seed orchards) the conditions should benefit flowering and fruiting. Selection of the appropriate site for the seed orchard will ease the management during and after the establishment.

Species Specific Requirements

— *Climate*: temperature, precipitation, photoperiod, wind exposure.
— *Soil*: texture, nutrients, pH, drainage
— *Topography*: flat/hilly/mountainous, exposure.
— *Altitude*: meter above sea level.

Knowledge of the species specific requirement and especially reproductive biology is crucial for proper matching of species and site, i.e. to be sure that the species will produce fruits at the site in question. The seed orchard should be established in the main portion of the species geographical and ecological range. For exotics, a seed orchard should be established only once the flowering and fruiting have been successfully proven in the area. The site should be selected in order to promote seed production only, unless it is combined with a genetic test. A site with average fertility is preferred to an area with high fertility since the latter may promote vegetative growth on the expense of reproduction. That may be especially important for low demanding species. A poor soil fertility can be manipulated with the application of fertiliser.

A flat or slightly undulating area is generally preferred since such area is generally homogeneous in soil structure and fertility. In a mountainous area, mountains will cast shade according to sun position (season and time of the day), that may have a crucial influence upon the reproductive biology of the trees. Special threats such as termites should be taken into consideration when selecting the area in areas where, e.g. termites are problems.

Management Requirements

— Accessibility
— Labour availability
— Close to administration
— Close to nursery
— Gentle topography

The seed orchard should be established as strategic for rational management as possible. Since the seed orchard is labour intensive both in terms of operation (establishment, maintenance, seed collection, etc.) and administration (record management, organisation of labour, etc.) a central place, e.g. near to a local forest administration office and nursery is preferable. Such areas may also be easier to protect. A flat or slightly hilly area, for example, abandoned agricultural land, is preferred e.g. because it permits mechanical operation by tractors and other machines.

Protection

— Strong winds, typhoons, etc.
— Fire
— Illegal logging/fuelwood collection
— Destroying animals

A seed orchard site should be established far from places where natural disasters are frequent. In areas where strong winds and typhoons occur, it should be placed as little exposed as possible, i.e. sheltered by hills or mountains. Fire is mainly caused by man and it is advisable that the orchard site be placed where e.g. grass fires are rare. In areas with severe wood shortage and consequent cutting and fuelwood collection, the orchards may be in danger unless the area has been negotiated with local authorities. Both wild and domestic animals can be destructive to seed orchards. Location close to national parks/game reserves or close to villages where for example, goats and cattle graze freely may be destructive. In general remote places should be avoided since they may be difficult to protect.

Isolation from Pollen Contamination

The seed orchard should be placed as far as possible to stands of the same species or species with which they can hybridise. If the species has a different phonology, crossing will be separated in time. Establishment of pollen dilution zones and, especially for wind pollinated species, a design with long axis of the orchard with the direction of the prevailing wind decreases the risk of pollen contamination. The seed orchard should provide optimal conditions for the pollination agent. In case of exotics with very specific pollination (e.g. birds or bats) pollination may be restricted due to lack or proper pollination agent.

Other Side Requirements

— Land tenure

— Possible future land requirement

The land tenure should be settled. Establishment of a seed orchard on community land may give problems with local community interests since it may interfere with possible traditional or alternative land use. Knowledge of possible plans for future alternative land use, e.g. roads, buildings, inclusion into urban zones, etc. is important for the longitivity of the seed orchard.

Spacing of the Trees

The trees in the seed orchard have a much wider spacing than those in normal plantations. For most species a final spacing, i.e. after rouging, of 10 meter is preferred. That corresponds to 100 trees per hectare. The initial spacing should be designed with the expected thinning in mind. If the trees are initially planted in 4 trees plot, with expected 3 of them plus may be 50% of the families to be culled, then a planting stock of 800 trees per hectare should be used, i.e. a spacing of 2.5 × 5 meter.

Site Preparation

The plants (either vegetative propagated or seedlings) are vulnerable to competition from weeds, to fire, diseases, browsing, termites, etc. during the first year of establishment. The better the site is prepared before planting, the easier the future management.

Clearing: Left woody material may attract pests, diseases, increase the risk of fire and impede mechanical management. The ideal one is a completely cleared area, free from trees, shrubs, stumps and big stones. That will ease the access to the area and mechanical operation. If the seed orchard is established on forest land, all trees and shrubs should be cut and removed or burned on the site.

Soil Preparation: Proper soil preparation improves the growth of the tree plants and promotes their competition with other plants. The site should be completely weeded before planting, e.g. by ploughing and harrowing. Mechanical cultivation improves soil structure, and consequently water holding capacity and draining. A hardpan impedes draining. It can be caused by tractors or if the soil has podzol profile. Hardpan is broken by deep ploughing or sub-soiling.

The soil should have the optimal acidity and nutrient level for the tree species to be planted. Soil tests are carried out prior to planting possible lime and fertiliser is applied before planting. Weed control can be facilitated by proper soil preparation. Several mechanical cultivations promote weed seed germination. Germinated seedlings are then destroyed by mechanical cultivation or by leaf herbicides. Soil herbicides can be used for difficult weeds.

Establishment of pollen dilution zones: Cleared areas are established between the seed stand and possible contaminating sources. The PDZ is a zone where trees of the same species and species that can hybridize with the species are removed. The wideness depends on the species and pollination pattern. The zones also serve as fire-breaks. For wind pollinated species a short vegetation of the PDZ is desired, usually 200-400 m, depending on species and wind exposure. If there is a prevailing wind during the pollination period, the width of the PDZ may be diminished from the other sides. For insect-pollinated species the width and the vegetation of the PDZ depends on the behaviour of the pollinator. Information should be sought on individual species.

Promotion of pollination: Promotion of the pollination is easiest for wind-pollinated species. Spacing is a method of improving the condition for wind pollination. The effect on insect-pollinated species is probably different from species to species according to pollinator, but details are poorly known. Promoting pollination by insects by putting up beehives is widely used in fruit orchards and agriculture and is applicable to seed orchards.

Demarcation of plots and rows: The orchard design is clearly demarcated. Replications. blocks and plots should be distinctly demarcated. The corners should be marked with conspicuous persistent poles / signs indicating necessary details. The material and way of putting up should be so that it will difficult to be removed or stolen.

Demarcation of pits and pitting: Follows the normal silvicultural practices. A final spacing in seed orchards of 10 meters is reasonable for most species in order to allow free crown development. Depending on the expected thinning the initial spacing may be 2-6 meters.

Planting: Normal silvicultural practices. Extra care should be taken with grafted and budded material since the grafting or budding site may still be vulnerable. Sometimes the root stock is planted in the field before grafting and consequently carried out in the field.

The planting usually involves several clones or families, and it is important that the planting is done according to a prescribed plan so that the identity of the individual tree is clear. Following measures should be taken in order to avoid confusion/mixing up of the plants:

— A plan of the orchard design should be brought to the field and used when planting spots are demarcated distinctly on the plan with the identity of the plant to be planted on the spot (clone/family - letter/number).

— The plants (or scions in field grafting) should be distinctly marked and kept separately.

— The plants are marked with a tag when planted. The tag carries a number / symbol of the plant similar to that of the map.

Beating up: Correct replacement of dead seedlings is crucial. If wilted, seedlings or grafted plants are replaced, and care should be taken that the replacement will not disturb the orchard design, i.e. plants should be replaced with the same clone / family.

Cover crops: A temporary legume cover crop may be planted between the tree plants. The cover crop is beneficial to the soil and may facilitate weed control and diminish erosion.

Orchard Design

The general requirement in the orchard design is:

— Minimise selfing

— Maximise outcrossing and mating of all genotypes.

— Simple and easy establishment and management.

— Allowing any number of clones or families to fit into the design.

The two most frequently used designs are complete randomised design and randomised complete block design:

— *Complete randomised design*: All available ramets of all clones or all families are distributed randomly all over the site.

— *Randomised complete blocks*: The area is divided into equally sized blocks which include one ramet of each clone or one member of each family. The ramets or families are located randomised within each block.

Randomised complete block design is the most frequently used if the design initially serves as a progeny test

Controlled Pollination

Controlled pollination can be conducted in which case pollen is collected from the desired male parent(s) and is applied to the flowers of the female parent(s). Controlled pollination may involve emasculation and isolation of the flowers of the female parents. If the orchard is designed entirely for hybridisation, planting of each clone in rows in preferred.

Silvicultural Management

Special care should be given to the plants during the first year of establishment during which they are especially vulnerable.

— *Weeding*: The plants should be kept from weed competition during establishment. Complete weeding should be done at least around each plant. Weeds between the plants may be cut or completely removed. If chemical weeding is applied, care should be taken in terms of time, weather condition, and doses of application (reference is made to the instruction of the individual herbicide).

— *Rouging*: Once the result of the progeny test is available, the undesired families or ramets are cut and removed.
— *Thinning and pruning*: In addition to rouging, the trees may be thinned and the branches systematically pruned in order to make an open crown with large flower production, and facilitate seed harvest.
— *Fertilisation and watering*: Water and fertiliser should be applied whenever necessary to give the trees optimal growth conditions. Conditions are especially important during flowering and fruit development. The amount and type of fertiliser differs from species to species.
— *Flower induction*: If flowering fails or is unsatisfactory, it may be induced or promoted by (1) imposing stress or (2) applying flower inducing hormones during flower differentiation.

Imposing Stress

Various methods of stress-imposing promote flowering:
— Water stress can be imposed if water availability can be manipulated by artificial watering or irrigation. A brief drought during floral differentiation will usually induce flowering. The watering should be re-continued after flowering.
— Hormone application: The active hormone in flower differentiation is gibberellin. Artificially manufactured versions of the hormone are sold under various trade names.

REFERENCES

Faulkner, R., "Seed Orchards", *Forestry Commission Bulletin No. 54*, London, 1975.

Granhof, J., "Seed Orchards", *Lecture note D-8*, Danida Forest Seed Centre, Humleback, Denmark, 1991.

Wright, J.W., *Introduction to Forest Genetics*, Academic Express, New York, 1976.

Zobel, B. and Talbert, J., *Applied Forest Tree Improvement*, John Wiley & Sons, 1986.

Bibliography

Andersson, E., "Seed stands and seed orchards in the breeding of conifers", *FAO/FORGEN*, 63-8/1, 1963.

Astinet, H., *Forest tree seed - specification of provenance*, Saatgut-Wirtschaft, World Seed Campaign Edition, 1961.

Barber, J. C. & Darby, S. P., "Georgia's seed certification program", *Journal of Forest*, 1959.

Bassett, M. J., *Breeding Vegetable Crops*, Avi Publishing, Westport, 1986.

Bewley, J. D. and M. Black, *Seeds: Physiology of Development and Germination*, Plenum Press, NY, 1985.

Bonner F.T., "Determining seed moisture in *Quercus*", *Seed Science and Technology*, 1974.

Byrum, J. R. and L. O. Copeland, "Variability in vigour testing of maize (Zea mays L.) seed", *Seed Science & Technolology*, 1995.

Ching T.M, Parker M.C., "Hydrogen peroxide for rapid viability tests of some coniferous tree seed", *Forest Science*, 1958.

Comin, D., *Onion Production*, Orange Judd Publishing, New York, 1946.

Cooke, D. A., and R. K. Scott (eds.), *The Sugar Beet Crop*, Chapman and Hall, New York, 1993.

Copeland, L. O., *Principles of Seed Science and Technology*, Burgess Publishing Co., Minneapolis, MN, 1976.

David Burton, "Undercooked Beans Can Cause Illness", University of Missouri, Retrieved on 2007.

Deno, N., *Seed Germination Theory and Practice* (2nd ed.), State College, PA, 1993.

Doijode, S. D., *Seed storage of horticultural crops*, Haworth Press, 2001.

Faulkner, R., "Seed Orchards", *Forestry Commission Bulletin No. 54*, London, 1975.

Fielding, W.J, Van Crowder, L., "Sweet potato weevils in Jamaica: acceptable pest?", *Journal of Sustainable Agriculture*, 1995

George, R. A. T., *Vegetable Seed Production*, Longman Press, Essex, 1985.

Gibbs, I.H., "Surveys of the sweet potato weevil *Euscepes postfasciatus* in Barbados during 1989 and 1990", *Proceedings of the Eight Annual Conference*, Grand Barbados West Indies, 1990.

Grabe, D. F. and D. Isely, "Seed storage in moisture-resistant packages" *Seed World* (Jan 24.) p. 4., 1969.

Granhof, J., "Seed Orchards", *Lecture note D-8*, Danida Forest Seed Centre, Humleback, Denmark, 1991.

Grogan, R. G., "Control of lettuce mosaic with virus-free seed", *Plant Disease Reporter*, 1981.

Harrington, J. F., "Drying, storing, and packaging seed to maintain germination, vigor" *Seedman's Digest*, 1960.

Hawkes, J.G., *The Potato: Evolution, Biodiversity & Genetic Resources*, Smithsonian Institution Press, Washington, D.C., 1990.

Hawthorn, L.R., Pollard, L., *Vegetable and Flower Seed Production*, The Blackiston Company, New York, 1954.

International Crop Improvement Association, "Minimum seed certification standards", Publ. No. 19. International Crop Improvement Association, 1959.

Isaac, L. A., "Problems and proposals for international certification of tree seed origin and stand quality", Proc. 5th World For. Congr., Seattle, U.S.A., 1960.

ISTA [International Seed Testing Association], "International Rules for Seed Testing", *Seed Science and Technology*, 1996.

________________, "Cold Test", *Handbook of Vigour Test Methods*, International Seed Testing Association, Zurich, Switzerland, 1987.

Jarvis, W.R., *Cucumber Diseases*, Agriculture Canada publication, 1992.

Johnson, R. T., J. T. Alexander, G. E. Rush, and G. R. Hawkes (eds.), *Advances in Sugar Beet Production*, Iowa State University Press, Ames, 1971.

Jones, H. A., and L. K. Mann, "Onions and Their Allies", *Botany, Cultivation, and Utilization*, Leonard Hill, New York, 1963.

Justin, O. L. and L. N. Bass, "Principles and Practices of Seed Storage", *Agriculture Handbook Number 506*, U.S.D.A. Agricultural Research Service, Washington, DC, 1978.

Koospol, *Seed and plants of forest tree species in Czechoslovakia*, Koospol, Prague, 1961.

Kozlowski, T. T., "Physiological implications in tree improvement", *FAO/FORGEN*, 1963.

Mabey, R., *Flora Britannica*, London: Chatto and Windus, 1997.

Matthews, J. D., "Factors affecting the production of seed by forest trees", *Forest Abstract.*, 24 (1): i-xiii, 1963.

McDonald, M. B., "Assessment of seed quality", *HortScience*, 1980.

McDonald, M., Copeland, L., *Seed Production Principles and Practices*, International Thomson Publishing, 1997.

Motes, J. E., "Pickling Cucumbers", *Production & Harvesting*, M.S.U. Cooperative Ext. Service Bulletin, 1977.

O'Sullivan, J., *Growing Pickling Cucumbers*, O.M.A.F. Factsheet, 1980.

Paged, *Forest tree seeds and trees*, Warsaw, Paged Foreign Trade Enterprise Export and Import, 1962.

Peet, M., "Practices and Materials Permitted in Some Organic Certification Programs", 2002.

Peralta, I. E. & Spooner, D. M., Granule-bound starch synthase (Gbssi) gene phylogeny of wild tomatoes (*Solanum* L. section *Lycopersicon* Mill. Wettst. Subsection *Lycopersicon*), *American Journal of Botany*, 1888–1902.

Piotto B., "Effects of temperature on germination of stratified seeds of three ash species", *Seed Science and Technology*, 1994.

Preston, S.R., "Checklist for use by potential donors before giving seeds to Pacific island countries after emergencies", *Phase II. PRAP Project 03. PRAP Leaflet No. 21*, Suva, Pacific Regional Agricultural Programme, 1999.

Rabinowitch, H. D., and J. L. Brewster (eds.), *Onions and Allied Crops*, CRC Press, Boca Raton, 1990.

Rose, F., *The Wild Flower Key*, London: Frederick Warne, 2006.

Ryder, E. J., *Leafy Salad Vegetables*, Avi Publishing, Westport, 1979.

Salaman, Redcliffe N., *The History and Social Influence of the Potato*, Cambridge University Press, 1989.

Seal, D. T., "Collection of cones from standing trees", *For. Comm. For. Rec. No. 39*. London, H.M.S.O, 1959.

Smith, A. F., *The Tomato in America*, Universi' of Illinois Press, 1994.

Stevenson, W.R., Loria, R., Franc, G.D., and Weingartner, D.P., *Compendium of Potato Diseases*, 2nd ed, Amer. Phytopathological Society, St. Paul, MN, 2001.

Tulstrup, N. P., *International trade in forest tree seed*, Unasylva, 13 (4): 196-201, 1959.

White, J.M., *Organic Vegetable Production*, UF/IFAS Coop. Ext. Serv., HS720, 1995.

Wright, J.W., *Introduction to Forest Genetics*, Academic Express, New York, 1976.

Zobel, B. J. et al., "Seed orchards - their concept and management", *Journal of Forest*, 1958.

Index